THE APPROACHING
GREAT TRANSFORMATION

THE APPROACHING GREAT TRANSFORMATION

Toward a Livable Post Carbon Economy

JOEL MAGNUSON

Seven Stories Press
New York

A Seven Stories Press First Edition

Seven Stories Press
140 Watts Street
New York, NY 10013
www.sevenstories.com

College professors may order examination copies of Seven Stories Press titles for a free six-month trial period. To order, visit www.sevenstories.com/textbook or send a fax on school letterhead to (212) 226-1411.

Book design by Jon Gilbert

Library of Congress Cataloging-in-Publication Data

Magnuson, Joel.
  The approaching great transformation : toward a livable post carbon economy / Joel Magnuson. -- A Seven Stories Press first edition.
     pages cm
  Includes bibliographical references and index.
  ISBN 978-1-60980-480-0 (pbk.)
  1. Sustainable development. 2. Environmental economics. 3. Economics--Sociological aspects. 4. Renewable natural resources. I. Title.
  HC79.E5M332 2013
  338.9'27--dc23
                              2012046142

Printed in the United States

9 8 7 6 5 4 3 2 1

To the memory of Christina 'Kit' Ward

I would like to thank my agent, Kit Ward, whose patience, support, and expertise was irreplaceable. Thanks also go to my editor, Gabe Espinal, and publisher, Dan Simon, and to my friends and colleagues who gave me encouragement, editorial advice, wisdom, and friendship: F. Robert Stuckey, Ivria Kaplowitz, Karen Deora, Karen Sanders, Marci Fickel, Rebecca Casanova, Kate Toswill and a special thanks to Natasha Gromova without whom this project would have never been possible. I am also indebted to those who helped me with my research by taking the time to talk with me about their businesses and projects.

—Joel Magnuson

CONTENTS

FOREWORD

by Helena Norberg-Hodge

Surrounded as we are by so many crises—social, economic, and environmental—it is easy to feel overwhelmed and disempowered. But there is good news too, particularly in a localization movement that is gaining strength across the world.

Most of the serious problems we face today have their roots in the globalization of economic activity, which has meant the continuous deregulation of trade and finance in favor of very big corporations and banks. The profit needs of these giant businesses lie behind the dramatic impoverishment of our natural and social wealth. In order to maintain "economic growth" in the global economy, the life support systems of our planet are being undermined and the fabric of community torn apart.

Despite what its proponents would have us believe, corporate globalization is not our only option. As we see in these pages, localization is not just an ideal, it is a concrete and viable alternative. More than that, it is essential for our survival: human well-being is intimately linked to the well-being of the planet, and localization protects both.

I first became aware of the impact of economic globalization in the early 1970s when I visited Ladakh, or "Little Tibet," in the Indian Himalayas. There, I found a remarkable culture, one based on cooperative and localized economic interactions. The people were well-fed and relatively prosperous, and possessed a *joie de vivre* I hadn't encountered anywhere in the industrialized world.

I was a linguist by training and—having fallen in love with the people and the place—I decided to stay in Ladakh to work on a Ladakhi-English dictionary. During the following years, I saw the region opened up to conventional development, making way for the intrusion of economic globalization. It was a rapid and disturbing shift. Subsidized food and other goods from the outside destroyed the market for local producers. The interdependent bonds that had held village communities together for centuries were replaced by competition and divisiveness. For the first time in Ladakh's history, there was unemployment and toxic pollution.

My book, *Ancient Futures*, along with a film by the same title, describes these changes in detail. Again and again I have been told by community leaders from diverse cultures all over the globe that "the story of Ladakh is our story too."

Worldwide, people are beginning to recognize the disastrous impact of the global economy on society and the natural world. From climate change to peak oil, from the epidemic of depression to unending wars over natural resources, from species extinction to the breakdown of community—all these problems are connected to the massive unregulated movement of goods and capital across the planet.

As this book makes very clear, an economy based so heavily on fast-disappearing sources of oil has no future. Ultimately, peak oil—in fact, peak *everything*—will force us to adapt our consumption-based lifestyle. But how will we adapt? The limits of growth are mathematically certain. As Magnuson says, "Resource depletion and climate change are the feet of clay of our economic system." More economic growth, more technology—in other words, more of the same—will not save us. But a different way of doing economics can.

Magnuson points out that stopgap measures will only make things worse: instead we need *systemic*—what he calls *institutional*—changes. Building a sustainable, just future requires reimagining our social and economic structures. Magnuson shows us how practical, realistic shifts in those structures can bring about a whole range of short- and long-term benefits.

I know from three decades of experience on five continents that locally-based economic structures are more transparent and accountable, and are better able to meet people's real needs. Shortening the distance between producer and consumer, particularly when it comes to the provision of the food we all need two or three times a day, is undoubtedly a necessary step forward.

I believe we also have to change policies at a governmental level. As things stand today, transnational corporations and banks are free to roam the globe in search of the most corporate-friendly working conditions, while at a national level they benefit from an array of tax breaks and subsidies that make smaller businesses seem "inefficient." For the sake of the planet's future, we urgently need to turn this around. Trade treaties need to be reformed, and the playing field leveled.

Changes are also needed in the values we—and especially our children—are taught through the media and advertising. As Magnuson writes, "People feel shame and alienation if they lack material goods." In this hyper-materialistic world of ours, we are encouraged to believe that happiness comes from owning the latest running shoes, the latest gadget. But this is a cruel lie. More than anything, our individual well-being depends on *connection*: on our sense of oneness with both other people and nature. It is precisely this sense of connection that the globalized consumer culture is breaking down. Recovering and nurturing it is perhaps our number one priority.

The Approaching Great Transformation is a timely call for a deeper understanding of our current predicament. A greater awareness would not only allow us to see that our crises stem from a fundamentally flawed global economy, it would also reveal what we need to do to turn things around. I call this *education as activism*, and believe it is the way to form powerful alliances for change.

Make no mistake, these alliances have the potential to make a real difference. Magnuson treats us to a cornucopia of new and emerging initiatives that are playing their part in the move toward the local: farmers markets and urban food gardens, LETS (Local Exchange Trading System, or Scheme) schemes and credit unions, and a range of community projects in places as far apart as Portland, Oregon, Totnes in the UK, and Mondragón in Spain. He tells us about the Transition Town movement and the thousands of "transitioneers" who are designing communities to prepare for peak oil. It's an uplifting read, giving us all much-needed hope for the years ahead.

Change will come; in the course of time, we will move in a different direction. The question Magnuson poses is key: Will this change happen because it is forced upon us or because we wake up and choose to live more creatively and kindly on the earth?

We have left things very late, but we know where we need to go. The future is ours, and we must claim it.

A pioneer of the localization movement, Helena Norberg-Hodge is the Director of the International Society for Ecology and Culture (ISEC) and Producer of the award-winning film, The Economics of Happiness.

ANOTHER SEA CHANGE

Every epic historical transformation began with economic turbulence and collapse. As a general rule, when the economic foundation of a society is significantly disrupted, it destabilizes everything else. Everything begins to change. The Roman Empire finally crumbled when its Mediterranean system of conquest, slavery, and trade was thrown into chaos by the Saracens who launched their invasions from Northern Africa. And the feudal system of the Middle Ages was brought down when its agrarian economy was ruined by plague epidemics and wars over religion. Today another historic sea change is beginning to take form and the underlying cause, again, is a major economic disruption. Though this time it will not be caused by invading and conquering armies or plague epidemics, the disruption will be a global decline in the availability of oil—the paramount and finite energy source that has supported all the major industrial systems for the last century. As resource limitation and climate change are beginning to shake the foundation of our entire way of life, another historic sea change is in the making.

Our current global system of production and commerce is driven by an imperative for endless growth and expansion. But limited supplies of oil and virtually all other resources are binding this expansion like a rope tightening around every economy in the world. Yet in a kind of blind and desperate attempt to keep charging forward with more economic growth, governments nearly everywhere are plunging themselves into massive and unsustainable debt. Whether it smashes into the wall of resource limitations and debt fatigue, or makes a shift away from business as usual, our global economy is going to change.

As we begin to feel these changes we will be forced to reexamine what we consider to be the good life. Many of us in the so-called developed world have high expectations for our career choices, our levels of income, the things we want to buy, and all the other accoutrements of a plentiful lifestyle. But this is all beginning to change. The cause of this change is a fairly simple but sobering reality: the resource base of our planet can no longer sustain it. When I refer to a declining resource base here, I am not only talking about oil, though oil is the big one. Other fossil fuels, water, topsoil, basic metals, and virtually every other resource available to us are being consumed to exhaustion. We often say that our desire for high material standards of living is to provide for our families—our children. The cruel irony is that our children and grandchildren will be the first generations to experience the full brunt of the economic and ecological damage we are doing now. That is, unless we do something now to change that.

In his book, *Collapse: How Societies Choose to Fail or Succeed* (2004, 2011), evolutionary psychologist Jared Diamond chronicles an impressive list of examples of how humans have a penchant for self-destruction. Though his examples are taken from different places and at different times in history, they all basically follow a consistent pattern in which entire societies collapse as a result of their destruction of the environmental habitat that would have otherwise sustained them. But we don't have to follow this pattern. We are an evolving species and we can learn from the lessons of history. We have conscious awareness and the ability to be proactive and possibly shape the direction of this historic sea change toward something better. This would be our lasting contribution—our great transformation.

The choices we make today as we adapt to mounting scarcity may be the most important historical events of the 21st century. We can choose to be forward thinking and work actively toward positive changes in a spirit of celebration and make this transformation in ways that are healthy, ecologically sound, economically stable, and just. Or, we can choose to be complacent, to continue treating our world as an infinite resource pipeline and an infinite waste dump, to brace ourselves for endless wars over resources, and to trudge through one debilitating crisis after another, pushing all of humanity through a long historical period of decline—a descent into a kind of Dark Ages of the third millennium. This would be a tragic downfall for humankind, but the ultimate cause of our downfall would not be the crises themselves. The cause would be our refusal to deal with the obvious fact that the Oil Age—the age of seemingly infinite abundance—is coming to an end.

The Twilight of the Oil Age

The Oil Age is basically the 20th century. Future historians, if there will be any, may consider it as one of those meta-historical epochs like the Stone Age or the Iron Age, though on the historical timeline it will be comparatively very short. On a chart of geological time, the Oil Age would appear as nothing more than a hair thin blip. Nonetheless, its significance in human history is profound.

Up to the end 19th century, industrialization and economic growth were powered by coal and the steam engine. Oil, or petroleum, was used mainly to refine into kerosene for lamps. That changed rapidly by the early 20th century. Coal was used to power external combustion steam engines, which use water boiled by an external heat source for power. Steam engines had their day, but soon were rendered obsolete as inefficient and cumbersome technology compared to the internal combustion engine powered by petroleum-based liquid fuels.

The Oil Age got its debut with the simultaneous development of two important technologies: the internal combustion engine and the commercial refining and production of oil. The internal combustion engine uses high-energy liquid fuels refined from oil, such as gasoline, that are combusted directly inside the engine for power. Like water, gasoline is lightweight, can be stored in a compact tank, and can easily be pumped directly into the engine. This allowed mechanical engineers to do away with the heavy baggage of the steam engine's pile of coal, boiler tank, and steam turbine. And while the design of the internal combustion engine was being perfected, the oil industry also quickly expanded its drilling and pumping operations and engineered hydrogen gas technology to efficiently refine or "crack" oil into gasoline. These simultaneous developments revolutionized the transportation and agricultural industries. The development of gas-powered cars, trucks, boats, aircraft, train locomotives, and tractors and other farm implements caused a productivity boom that led to subsequent waves of economic growth and development. The ascent of the Oil Age was all but preordained.

Henry Ford's company mass produced about 250,000 cars in 1913, and by 1929 the number jumped to 23 million. About the same time the Wright brothers developed various models of the Wright Flyer aircraft also powered by internal combustion motors. Between 1892 and 1903, gasoline powered tractors were also being developed, and by 1917 Henry Ford began mass producing the Fordson tractor. Tractors replaced oxen, horses, and human labor in the fields and food became much more abundant and much less expensive. The pace of economic production and daily life accelerated dramatically, and when the jet aircraft engine was developed in 1945 the world suddenly became a much smaller place. In the decades that followed, auto manufacturing, industrial agriculture, and air travel became leading industries and they were all based on the steady and abundant supply of cheap oil.

Cheap oil has been more than the basis for fuels to power industrial growth. As engineers in the 1960s developed chemical fertilizers and pesticides synthesized from oil, agricultural output experienced another major productivity boom known as the Green Revolution. The technology was applied everywhere, but its impact was felt particularly in developing countries that were experiencing severe problems of hunger. As a result of applying synthetic chemicals to wheat production, wheat output increased in developing countries from 750 kilograms per hectare in 1950 to well over 2,500 kilograms per hectare in 2000. Global food production increased exponentially, but the Green Revolution rendered much of the world's population dependent on oil, mechanization, and prodigious amounts of irrigation for survival.

By the 1970s, however, a global oil crisis and an emerging environmental consciousness prompted concerns about becoming overly dependent on oil. Physicist and mathematician Albert Bartlett issued grave warnings about allowing our entire economic system, including food production, to become so heavily reliant on oil. Among the many notable aphorisms by Bartlett is this one from his 1978 article, "Forgotten Fundamentals of the Energy Crisis": "Modern agriculture is the use of land to convert petroleum into food."[1] Bartlett cautioned that the approaching decline in oil would seriously impair food production. He warned, "It is clear that agriculture as we know it will experience major changes within the life expectancy of most of us, and with these changes should come a further major deterioration of world-wide levels of nutrition." In other words, Bartlett was telling us that over the last several decades shortages in our global food supply have been temporarily and artificially kept at bay with petrochemicals. His warnings were largely ignored, but now they are coming back to haunt us.

About the same time, David Pimentel of Cornell University issued the following warning:

As a result of overpopulation and resource limitations, the world is fast losing its capacity to feed itself . . . More alarming is the fact that while the world population doubled its numbers in about 30 years the world doubled its energy consumption within the past decade. Moreover, the use of energy in food production has been increasing faster than its use in many other sectors of the economy.[2]

His warnings were largely ignored as well, and since then our use of oil has not slowed down. In fact it has increased.

Oil has become the very foundation of the global economy. Although the bulk of it is used for transportation fuel, it still supports multiple layers of industry. In the transportation industry, oil is not only used to move goods around the planet, it is also, by association, used to sustain car and truck manufacturing, auto parts and repair industries, tire manufacturing and tire shops, auto insurance, interstate freeway construction, and the list goes on. Oil is the primary ingredient for making other products including plastics, synthetic fibers, cosmetics, and again, food. These industries that are either directly or indirectly dependent on oil constitute the very spine, heart, and lifeblood of the entire global economic system. They spawned massive flagship corporations like Exxon, Ford Motor Company, and General Motors. They also created millions of steady, well-paying jobs that contributed to the ascent of a solid middle class and the suburbanization of American communities, which further entrenched our oil dependency. None of this has really changed and oil remains a paramount and ubiquitous resource, and there is nothing in the universe to replace it.

As the Oil Age comes to an end, these industries will begin to fade and it's unclear what will follow. It's unclear because where we go from here will depend on the choices we make and the steps we take now, and those choices and steps are yet to be determined. I emphasize here and now because we are currently in the peak oil moment. That is, the supply of oil, this amazing finite and one-time resource is now making its inevitable turn from growth to decline. That puts us on a veritable cusp of a historical transformation.

A Brief Update on Peaking and Warming

Much has been written on peak oil and global warming over the last several years, yet virtually nothing beyond symbolic gestures has been done about them. It is not necessary to rehash another painful litany of the world's environmental woes as we have become quite familiar with the problems of melting ice caps and receding glaciers, species extinction, unstable climate, and resource depletion. The data keep streaming in and serve only to reinforce the warnings that scientists and analysts have been issuing for decades. Twenty years ago leaders from around the world met in Rio de Janeiro and pledged to protect and restore the health of the planet's ecosystem. In this effort we have failed miserably as conditions have worsened and the rate of resource depletion has accelerated. As long as we cling to business as usual, these problems are not going to go away and, painful as it may be, it is important for us to keep the dialog alive. It would be unwise of us to raise the discourse on oil depletion only when another horrific event occurs like another massive oil spill, nuclear power plant meltdown, or the damage and fury unleashed by Hurricane Sandy.

Kenneth Deffeyes, geology professor at Princeton University and oil industry expert, nominated Thanksgiving Day, November 24, 2005, as Peak Oil Day.[3] His nomination came with the caveat that the peak is more like a plateau that will sustain a series of peaks before showing a clear and definite descent. As of this writing, the most recent Energy Information Administration (EIA) data shows global oil production reached its highest peak in October 2010, at about 86.8 million barrels per day. But this is only slightly higher than the previous peak of 86.7 million in July 2008. Since 2005, global production has been suspended in a peak plateau ranging between 83 and 87 million barrels per day. Moreover, geologists have made it clear that in 2005 conventional oil production had peaked.[4] Conventional oil is the stuff that basically just squirts out of the ground. So most of any new oil resources will be cooked out of tar sand, mined from under porous rock, or will have to be pumped from miles beneath ocean and sea beds. This makes oil increasingly hard to get and therefore costly.

Aside from EIA data, there are other ways to see that global oil production is peaking. One is the approximate forty-year lag between oil discovery and oil production. Global oil discovery peaked in 1965. By this measure, global oil production would be peaking on Deffeyes's Peak Oil Day. Further evidence of peak oil can be gleaned from the financial decisions made by the large oil companies. As with any other industry that reaches the climax of its expansion, the oil industry has fallen into a cannibalistic merger wave. Feeling the relentless pressure from their shareholders, they are devouring the oil reserves of other companies through a series of corporate mergers so that they can boast of steady future revenues. As oil companies find it impossible to find new oil reserves, their share prices begin to fall. Fruitless exploration, massive capital requirements for offshore drilling, and small returns are putting oil company stocks at risk. The companies have been using the massive cash they have accumulated to buy back outstanding shares in a desperate attempt to boost their share prices.

When oil companies try drilling and pumping oil from 4 miles beneath the ocean, or cook it from tar sands, this is pretty good evidence that the easy-to-get oil is close to depletion. Standard economic theory argues that as oil prices rise due to scarcity, it becomes cost effective to develop less conventional ways of extracting and processing oil. This is true to an extent, but you can only take this argument so far because unconventional oil production has finite limitations. Geologists predict that those sources would be unlikely to produce more than 4 or 5 million barrels per day, a small fraction of the total demand. While world production remains currently suspended between 83 and 87 million barrels per day, consumption is expected to climb to as high as 120 million barrels a day. This implies that we are not only using up current production, we must also deplete current reserves and inventories just to keep up with demand.

Critics of the peak oil argument will want to squabble over this or that detail about when oil production will or did actually peak. But pinpointing the exact month of peak production is largely a trivial academic argument. Whether we call it a peak or plateau depends on how closely we choose to look. Since we are looking at this in a broader historical context, we'll take the longer view. From that view the moment of peak oil is now. How we define "now" could be last year, today, or five years from now, but all of that will be decided by future historians. What is more important is that the upshot of all of this is the critical inflection toward energy descent. Geologists anticipate that by 2040, US oil supplies will have descended to 90 percent below its production peak that occurred in 1970, and world oil production will have descended to at least 63 percent below its current production peak. There is no question that oil will get scarcer and much more expensive. It is a finite resource and we don't have to wait until it hits $300 per barrel to draw that conclusion. As the price of oil is a benchmark for energy prices overall, energy in general is going to get much more expensive. The higher the price of energy, the greater the shock waves it will send through the global economy and therefore the more significant disruption it will be.

If we slide into energy descent while continuing with business as usual, nations everywhere will embark on a mad scramble to find other sources of energy to power economic growth. The most likely scenario will be to ramp up the use of the more abundant fossils: coal and natural gas. But as we increase our consumption of these fuels, we will truncate their lifespans and cause them to peak within a few decades or sooner. Energy-hungry eyes have also turned, once again, to nuclear power despite the ominous disaster at the Fukushima Daiichi nuclear facility in Japan. Like turning into another dead end in a maze, ramping up nuclear power will only hasten the condition of peak uranium. The pollution created by burning coal and gas and the dangers associated with nuclear power will also increase proportionally. And once we have finished with all this madness, we will end up right back where we started: facing energy descent. Only by then the world will be a far more toxic and dangerous place to live.

In his 2008 campaign for the presidency, Barack Obama talked a great deal about "clean coal." This is another term for carbon, capture, and storage (CCS) processes or "coal scrubbing." Clean coal has been championed by both the coal industry and by some environmental organizations as a means by which carbon emissions can be reduced from coal-burning power plants. The process involves separating out carbon from the plant's effluents and then pumping the sequestered carbon deep into the crust of the earth where it can be reabsorbed. This is almost literally a process of sweeping the problem of carbon pollution under the rug. Obama spoke of government plans to help the industry develop this technology and to mitigate the effects of global warming. But shortly into its first term, the administration abandoned the idea because of the enormous expense that would be involved compared to the minimal benefits it could provide. Another concern raised by the critics of clean coal is that with carbon scrubbing and government subsidization we cement our dependency on coal for energy.

Because of its tremendous importance oil is like the barometer or leading indicator for the depletion of resources in general. What happens to oil basically happens to all resources. To keep track of our use of a broader spectrum of our resource base, economists do what they call "material flows analysis." This is basically the process of tracking and measuring the flows of actual material things that go with economic production and distribution. This includes the final products, but also includes iron, copper, aluminum, wood fiber, topsoil, fossil fuels, vegetation, fresh water, and pretty much everything else that is used to make things. Though our technology is supposed to be making us more efficient, the material flows analysis data shows that the rate with which we burn through materials has increased despite the wonders of technology.

Throughout the years between 1980 and 2005 there was a general assumption that technology was going to transform economic production in ways that would require fewer resources. This included the so-called "New Economy" years of the 1990s in which the US economy was supposed to be shifting to a resource-light "information economy." Computers and the internet were supposed to make us efficient and help us save on material resources, but the result was the opposite. During that twenty-five-year period, the consumption of materials in North America increased by 54 percent: from 6.6 to 10.1 billion metric tons. As this region's population only increased by 35 percent during those same years, it is clear that material consumption per person has steadily risen. Information technology served to pick up the pace of globalization and economic expansion such that consumers were able to wolf down pieces of our planet in unusually large chunks. As we continue to do this we have charged headlong into a condition that Richard Heinberg succinctly calls, "peak everything" (*Peak Everything: Waking Up to the Century of Declines*, 2010). And this would include peak water.

Fresh water is in peril. In most places in the world fresh ground water usage is at its maximum. Though it is a renewable resource, we are using or damaging it at a rate faster than it can be replenished. Most of the water is used for food, drinking, and sanitation for an ever-enlarging population of consumers. Water tables are also falling at an alarming rate. In China and parts of India, the world's most populous regions, the water tables are dropping at a rate of about three feet per year. The water table in the vicinity of Mexico City, again an area of large human population density, has dropped by over seventy feet in fifty years, and in Saudi Arabia, aquifers are down 50 percent from their levels in the 1990s. Saudis, particularly in the east, are struggling with their difficulties in growing food because of scarce water. In response they are turning more and more to desalinization plants to create more freshwater. Their Shuaibah 3 desalination plant is the largest and most technologically advanced in the world, and it is also extremely energy intensive. Saudi Arabia depletes its fresh water supply by about 700 billion cubic feet every year, and in order to compensate for this with desalinized water, it would take about 300,000 barrels of oil or the equivalent in natural gas. Along with energy descent so goes the desalinization option of obtaining fresh water.[5]

As water tables are dropping and aquifers are being depleted in the US, complex and intractable problems are surfacing in multiple sectors of the economy. Much of the nation's water supply is supported by public utilities that rely on municipal bonds for credit. As water shortages become increasingly severe, conflicts over water usage arise. Because public utilities have to pay their own legal fees, they become vulnerable to legal battles over water supplies. This has spurred credit rating agencies to downgrade the municipal bonds in areas where water shortages are becoming chronic problems. One of the most stable and risk-free investments, municipal bonds, is now coming under attack as water gets scarcer. In other sectors high tech companies such as semiconductor manufacturers are rapacious consumers of fresh water and are being pitted against farmers for water rights. Farmers are also pitted against developers, and developers against sports fishermen, and so on. These conflicts will get much worse as drought conditions in some areas force human migration to other areas where water supplies are still available.

Water depletion is only part of the story. Industrial, chemical-based agriculture is destroying more than 10 million hectares, or 23 billion tons, of arable land each year as the rate of topsoil ruination exceeds the rate at which soil can be regenerated. Industrial agriculture is also damaging water supplies with contamination. Mercury, fertilizer runoff, industrial chemicals, sewage and human waste, animal fecal matter, storm drains, mining chemicals, rust inhibitors, heavy metals, pesticides, herbicides—all contribute to the damage and contamination of fresh water. Natural processes can and do clean water supplies by breaking down waste material and chemicals and returning them to the crust of the earth. But these processes, like the processes of hydrologic cycle of replenishment, are very slow compared to the rate at which we damage and pollute.

Some gains can be made with water conservation efforts and efficient usage and water recycling. But these gains are being eclipsed by growing demand from population increase and by damage done by industry. Like peak oil, the easy-to-get water supplies are already being taken and what is left is the hard-to-get water, or no water at all. To get a sense of magnitude of this problem, Sandra Postel wrote in *Last Oasis* (1997):

> . . . if 40 percent of the water required to produce an acceptable diet
> for the 2.4 billion people expected to be added to the planet over the
> next 30 years has to come from irrigation, agricultural water supplies

would have to expand by about 1,750 cubic kilometers—equivalent to roughly 20 Nile rivers, or 97 Colorado rivers. It is not at all clear where this water is to come from.[6]

Aside from the loss of fresh water and arable soil, another problem affecting our food supply is climate change. In 2010 and 2011, agricultural regions around the world have been experiencing extreme irregularities. Blistering temperatures in Russia resulted in drought while the normally drier wheat producing areas of Saskatchewan and Australia were pummeled with dangerously heavy rainfall, and close to half of China's winter wheat crop has been crushed by an unusual dry spell. When farmers can't produce, scarcity follows and prices take off.

Prices of basic food staple items like wheat and corn have soared to record levels. According to prices set by commodities markets in Chicago, wheat rose by 74 percent in 2010 and corn prices increased by 87 percent in 2011. The world's poorest families are the most severely affected by high food prices as they typically spend over half of their meager incomes on food. With oil prices climbing back over $100 per barrel, this will only worsen the situation for the poor. One of the most at-risk places in the world is the arid region stretching from sub-Saharan Africa to Northern Africa and to the Middle East. Food riots and political protests quickly spread to Algeria, Yemen, Bahrain, and then to the deadly fight in Libya. The political instability that followed may come to be seen as the first large-scale and undeniable social crisis stemming from global warming.

That these things are happening should not come as a great surprise. The Intergovernmental Panel on Climate Change (IPCC) has issued a number of warnings that this would happen with rising temperatures and that we would have to deal with the global warming crisis on multiple fronts. The IPCC reports outline a possible sequence of events such as seawater levels rising by anywhere between seven and twenty-three inches and flooding of coastal regions including many highly populated urban areas, changing weather patterns, warmer ocean temperatures, arctic conditions developing in some temperate climates, drying trends, floods, falling crop yields, and drought. Government institutions are already making preparations for dealing with an onslaught of problems that will unfold as a result of changing climate conditions.[7] The US Department of Defense has raised concerns that these could easily develop into national security problems as they escalate into food riots, mass population migrations, and eventually violent conflict.

The evidence that climate change is caused by global warming and that it is caused by human emissions of carbon dioxide into the atmosphere as presented by the IPCC is abundant and clear. The science of it really isn't that complicated. Given the earth's distance from the sun, it should be colder than it is. The reason the planet can stay warmer is that its own atmosphere is like a blanket of insulation trapping heat that has made the long journey from the sun. The elements in that insulation are greenhouse gases, principally carbon dioxide. As solar heat enters the atmosphere, the planet absorbs some of that energy like a sponge and the rest is reflected back out into space as infrared radiation. Carbon dioxide traps that infrared heat and prevents it from leaving the atmosphere. The more carbon dioxide we release into the air, the thicker that blanket of insulation becomes, and the planet gets increasingly warmer.

What is more complicated is trying to discern the impact warming ambient temperatures will have on our climate, weather patterns, and ocean levels. Nonetheless, in March 2009, the scientists who gathered in Copenhagen, Denmark, for the UN Conference on Climate Change issued their "Synthesis Report—Climate Change: Global Risks, Challenges and Decisions." In their report, they drew the following conclusion:

> The climate system is already moving beyond the patterns of natural variability within which our society and economy have developed and thrived. These parameters include global mean surface temperature, sea-level rise, ocean and ice sheet dynamics, ocean acidification, and extreme climatic events. There is a significant risk that many of the trends will accelerate, leading to an increasing risk of abrupt or irreversible climatic shifts.

Sadly, the issue has been politicized and mangled by the political right wing of the US Republican Party. Their refusal to acknowledge the existence of global warming is absurd theater and blocks our ability to have sound public discourse on the subject—though there are signs that the political mangling might be changing. In December 2011, the United Nations conference in Durban, South Africa, put forward the "Durban Platform" which, if ratified by a UN quorum, will create an international legal agreement to reduce carbon emissions and to begin implementation in 2020. Just beneath the surface of climate change politics, however, are the powerful oil and coal industry lobbyists who are framing the discourse—arbitrarily creating controversy to keep the issue trapped in a debate rather than achieving a consensus. But the issue of global warming runs much deeper than national politics.

In order to raise awareness about climate change, the award-winning documentary, *An Inconvenient Truth*, directed by David Guggenheim, was released in the spring of 2006. The film is based on the work of former Vice President Al Gore as he sought to draw attention to the harsh realities of global warming. The film and Gore's companion book were enormously popular and seemingly overnight global warming shifted in the popular media from an obscure science topic mired in controversy to an alarming scientific reality. Widespread concern about global warming was felt in Washington and prompted the federal government to respond.

Yet only a couple of years after the movie was released the topic of global warming was pushed aside to make room for the media's new obsession—the economic recession—and global warming all but vanished from the public eye. The need to restore the economy to growth clearly became the nation's top priority and all other concerns were ordered to the back of the bus. This was not a coincidence or a random event; this is a recurring theme. The needs of the economic system for growth and expansion always takes a front seat, while environmental concerns consistently remain marginalized as controversial or ephemeral.

Meanwhile greenhouse gases continue to build in the earth's atmosphere and are rising above worst case scenario projections. The US Department of Energy released a report in October 2011, that indicates the heat-trapping gases jumped by a record amount in 2010 to levels much higher than those forecast by climate experts just a few years earlier.[8] Even though awareness of climate change has been firmly established, the global economy spewed an additional 512 metric tons of carbon over 2009 levels. As always, the United States and China, the world's two largest economies, topped the polluters list as they accounted for more than half of the additional emissions.

Our Great Transformation

The twin crises of resource depletion and climate change are the feet of clay of our economic system. Sustained disruptions in economic production and distribution will eventually destabilize the system and, as we know from history, this will precipitate a much larger sea change. An economy that is driven by powerful corporate and government institutions to continue growing will eventually hit the wall of physical limitations. And the most significant physical limitation is the decline of oil. As the Oil Age comes to its inevitable end, our habitual ways of behaving economically and ecologically will undergo a transformation.

The words "economy" and "ecology" both have a prefix derived from the classical Greek, *oecos,* meaning "household." This common linguistic origin implies that how we pursue our livelihoods and how we interact with our natural environment are one and the same. Economic activity will always have an environmental impact and environmental changes will always alter the conditions of our economy. For decades scientists have been warning that both of these are going to change. Either the threat of climate hell that will arise from global warming will compel us to change our economic habits, or changes in our habits will be forced on us by energy descent. Or perhaps it will be a combination of both. However it plays out, large scale changes are on their way.

Much attention has already been paid to the twin problems of peak oil and climate change. Mountains of articles, books, and documentaries have been published on these subjects. From this work it appears that a consensus is forming around what investment banker and oil industry analyst Jeff Rubin said: that for each of us our world is about to get a whole lot smaller (*Why Your World is About to Get a Whole Lot Smaller: Oil and the End of Globalization*, 2009). All indications show that the growth-driven, globalized economic system that is powered by fossil fuels is finally hitting the wall and that this will force us to redirect our attention to economic re-localization. And a positive adaptation to changing resources and environmental conditions hinges on our ability to develop vibrant local economic alternatives that are ecologically sound and more democratically governed.

I agree with this general conclusion, but I must also point out these conclusions are nothing new. For decades people have been talking about going local and going green as an alternative to the global economic juggernaut. Farmers markets and efforts to produce sustainable or environmentally friendly products go back to the 1960s and today organics have grown into a multi-billion dollar industry. Yet our track record on resource depletion and carbon pollution has indicated little improvement; in fact, if the Energy Department's report is a reliable indicator, it is getting worse. Part of the reason little is changing is that what was once considered "green" or "sustainable" has been co-opted and is looking more conventional all the time. It is now an all-too-familiar story of what Heather Rogers calls "green gone wrong" (*Green Gone Wrong: How Our*

Economy Is Undermining the Environmental Revolution, 2010)—that going green is more of a marketing ploy to tap into the pricier organic or fair trade markets. And the cheering squad for the "going local" movement boasts how local small businesses are growing, generating profits, and expanding markets. Yet these are the standard measures of success according to conventional business models, and are the same measures of success that compel businesses to burn through more energy and to find more consumers with money to spend.

Granted, such efforts to go green and local may lighten the burden we place on our resources, but they are ultimately driven by the same conviction that profit, growth, and fiduciary responsibility to shareholders are of paramount importance. As such, this overrides virtually every genuine attempt to take a different path and, at best, leaves us with half-hearted attempts at finding real solutions.

Others still argue that the solutions are to be found in new technology and government policies. The widespread belief in a technological answer to environmental problems has been elevated to a kind of secular religion. It instills a sense of blind faith and serves as a comforting palliative for those who choose not to look too closely. "Don't worry," as the familiar old chestnut goes, "if we can put a man on the moon . . ." Yet the irony of our technological sophistication is that it has accelerated the pace at which we are depleting resources. The more efficient we become at producing and consuming things, the less costly they become and therefore the more we produce and consume.

And when we look to federal government policies for solutions, we are sadly barking up the wrong tree. In September 2011, President Barack Obama issued an executive order to not enforce the Ozone National Ambient Air Quality Standards set in 2008. Obama's reason was the same as that of the previous George W. Bush administration's refusal to join the international efforts to reduce carbon emissions—it's bad for the economy. The federal government is reluctant to pass or enforce any legislation that could genuinely address the problems of environmental damage or resource depletion because, to be more precise, it is bad for business as usual.

In my view, the seemingly intractable problem in all these situations is institutional. I see nothing inherently wrong with going local, organic, or green; nor is there anything inherently wrong with trying to create better technology or government policy. What renders virtually all of these efforts nonsolutions or half measures is that they are built on an assumption that the same business and government institutions that brought us to this point of resource exhaustion and climate horror are somehow going to solve the problems. This is as contradictory and self-defeating as trusting a class of slave owners to make something humanitarian out of the institution of slavery. Large corporate and government institutions are just not equipped to deal with these problems, programmed as they are to build their empires and conquer and such.

The central message of this book is this: if the problem is institutional, then institutional transformation has to be the solution. As we will see in a later chapter, institutions are social structures that script social behavior with rules, norms, and shared strategies. If we seek to truly adapt to the profound changes that will accompany the end of the Oil Age, we will have no choice but to create new social structures that will script our social behavior with entirely different rules, norms, and codes. If done with insight, wisdom, and a wholesome set of beliefs about our relationships with nature and each other, then this will be our great transformation. Citizens can develop new and very different economic institutions that are centered not on endless growth and expansion, but on self-reliance, ecological permanence, stability, and a celebration of human creativity. In time these institutional developments will help guide us to the positive evolution of economic systems and human culture. This will also require a transformation of our mindset, expectations, and what we consider to be the good life. This is a tall order, but, as Paul Rogat Loeb reminds us, "the impossible will take a little while" (*The Impossible Will Take a Little While: A Citizen's Guide to Hope in a Time of Fear*, 2004). And, in fact, the process has already begun.

The groundwork for this great transformation is being laid by the transition movement. In "Chapter One: From Transition to Transformation," I tell the story of the transition movement with some background of the movement and its philosophical underpinnings. Its main goal is to foster community-based projects called "transition initiatives." These initiatives are structured around a transition model for community action developed by Rob Hopkins and other transition pioneers. Most of the ingredients in the model are directed toward creating vision, raising awareness, organizing, training and reskilling, democratic participation, and planning. The most substantive part of the model is a community-centered Energy Descent Action Plan (EDAP) for reducing fossil fuel consumption and carbon emissions. By their own admission, however, Hopkins and others who developed the model acknowledge the model needs to be developed further with a more comprehensive vision of economic systems and how they function. They have left it an open-ended questio as to what kinds of business models and economic institutions need to be developed to implement the EDAP and complete the transition to a whole, stable, resilient, post carbon local economy. But this is not a minor question. In fact, it is perhaps the biggest and most difficult question because successfully making this transition will require more than just coming up with new business models or technologies. That is, effectively dealing with problems of global warming and peak oil will require more than a transition, it will require a transformation.

In "Chapter Two: Economies in Transition—A Systems View" I flesh out a broader understanding of economics by taking a systems or holistic view. Observing the actions taken by a particular business or choices made by individuals in isolation is not enough to grasp how we have arrived where we are or what we can do about it. A genuine understanding of economic transformation requires a broader, systems view of the interconnectedness of people, economic institutions, and their habitats. It also requires an understanding of the importance of culture. Consumer culture and the choices we make in it are as much driven by our economic system and the corporate institutions that dominate it as by choices individual make within it. I encourage citizens to become active agents for change by transforming themselves inwardly and transforming their communities outwardly.

Though often what we mean by change is poorly understood. "Chapter Three: This Road Paved With Good Intentions" is more of a cautionary tale about our attitudes toward economic growth, money, and our expectations regarding trying to live in a truly sustainable way. Growth for growth's sake is a systemic condition driven by the financial sector's need to maximize returns to their investors, which creates a widespread preoccupation with sales and market expansion. It has become deeply ingrained in American culture where there is a widespread expectation that financial investments will keep growing forever. For this to happen, the production and consumption of goods must grow forever as well, but it cannot because the carrying capacity of the planet won't allow it. Many of us are in denial about this, particularly economists, and have chosen to believe in a certain fiction. The fiction is that we can not only pursue both ongoing growth and sustainability, but that benefits of both can be shared among everyone equitably. I refer to this as "green economics." If we are serious about ecological permanence and social equity, then green economics needs a sober and critical examination. I present a number of caveats regarding "going green" and "going local." Up to now the movements toward "local" or "sustainable" or "organic" have been profitable, but have not moved our economy toward any measure of ecological permanence or equity because they originate in conventional institutions. Again, to move toward ecological permanence would require real change, not half measures—a much deeper transformation and a clean break from business as usual.

To make such a deep transformation will require some vision and inspiration. For this, in "Chapter Four: E. F. Schumacher and the Meta-Economists" I revisit the work of visionaries who for over a century and for various reasons have been making appeals for the very same kind of economic and cultural transformation. It is helpful for us to see the work of critics and visionaries who were bold and idealistic enough to share their vision of what such real changes might actually look like. This exploration will take the reader through the works of John Ruskin and William Morris, who sought to create a new intellectual force and political agenda through a revival of craft traditions; Patrick Geddes's ideal of "civic regeneration"; Ebenezer Howard's vision of "garden cities"; Randolph Bourne's conception of "beloved community"; rich contributions by Lewis Mumford and the Young Americans who envisioned organic communities. They are meta-economists because they are seeing economics as something that should serve higher moral and ethical values that transcend the specifics of economic activity.

This chapter concludes with an overview of E. F. Schumacher's work on economics and philosophy. He criticized standard economic theory for not only ignoring these problems, but actually fostering destructive attitudes by rationalizing greed and selfishness, and instilling a certain callousness to fragile ecological systems. He was particularly concerned with developing countries following Western economic models of development and instead argued for grassroots models that shift the locus of economic activity away from large corporate and national systems toward smaller, semi-autonomous systems supported by appropriate forms of technology. As a philosopher, Schumacher argued that economic activity should be guided by wisdom rather than greed, and carried out with the aim of ecological permanence rather than exploitation. He emphasized the development of the individual person through good work, wisdom found in all the great spiritual traditions, and a reverence for nature.

What is absent from both the transition literature and the works of the meta-economists is a specific discussion on how to craft new institutions. The first part of "Chapter Five: The New Monastics" is specifically about building new institutions. For this I draw on the model of institutional diversity developed by Elinor Ostrom. The basic notion is that the prescribed behavior of our established institutions has brought us to the brink of social and ecological ruin and therefore those prescriptions will have to change. The rest of the chapter is a survey of unconventional practices and business models. The idea here is to tell their stories, particularly those about what is working well, and to see their practices as the substance from which new institutions can be crafted. These models may not have a perfect scorecard, but they represent practices that can help lead us in a new direction.

My evolving list of subjects is made up of people in contemporary society who are rejecting the centralized corporate system and creating their own alternatives without government mandates, basing their work on alternative values and philosophies. I look at innovative business models such as the "B Corporation"; activists in Pennsylvania who have started movements directed at revitalizing local citizenship; a brewery in Colorado that is employee owned, wind-powered, and converts gray water into electricity; an employee-owned solar power business in Colorado; city planners in Oregon who have for decades been redesigning the Portland metropolitan area specifically to reduce dependence on fossil fuels and in the process have created a mecca for bicycling and transit enthusiasts from around the world; a network consisting of a permaculture credit union, farmers market institute, and land trusts to promote local food production in New Mexico; a network of urban cooperatives in Ohio committed to local economic development; and other examples of work consistent with the values of livability, social and ecological stability, and equitability. As we look at all these cases together, we can begin to develop a vision of an institutional "dream team" in which these institutions cohere into a system—an alternative mutual support network.

Finally, *The Approaching Great Transformation* concludes with "Chapter Six: Education and Our Great Transformation." Transforming our economic system into something new and better will require the development of educational programs aligned with this effort. Moving toward a livable future will require that people develop new skills and technologies, create new cultures, foster creativity, and raise awareness. Educational institutions need to develop new curricula and workshops that will provide tools and guidelines for people who seek to learn how to become citizens in their communities and to design, heal, manufacture, grow, repair, plan, rebuild, as well as to become the visionaries of their own generations.

CHAPTER ONE

FROM TRANSITION TO TRANSFORMATION

Though it is going to take some time, the transformation to a new economic system is underway. This is happening in part because all the major economic systems of the world are mired in economic stagnation, buried in massive debt, and plagued with chronic high unemployment. Our economic systems and the institutions that comprise them are incapable of dealing with the problem of limits to growth. Leaders in government and the central banks are tossing around trillions in cash in one stimulus or bailout plan after another, hoping to prop up the illusion that our economies have stabilized and can continue to grow for all time. But this is getting harder to do as bailout fatigue is setting in among populations around the world. Meanwhile climate is becoming more erratic and economies are becoming increasingly unstable.

These crises are forcing economic change though it remains uncertain if they will change in ways that are good for people or the planet. Given this uncertainty and a belief that our economies can change for the better, activists everywhere are attempting to make their communities more resilient and to adapt to these crises in positive ways. The transition movement, though still evolving, is arguably the most focused, community-centered response to peak oil, climate change, and the increasing volatility of the global economy. The movement is an international network of people and organizations working proactively to help citizens undertake projects in their communities that will scale down energy consumption, do less environmental damage, and make them more economically self-reliant. However, we will see that adapting to a new life at the end of the Oil Age will be a massive undertaking and success will require deep transformation of our basic economic institutions as well as our own consciousness.

The Origins of the Transition Movement

The transition movement is relatively new. Its first initiative, Transition Town Totnes, was unleashed in Devon, UK, in September 2006, by Rob Hopkins, Naresh Giangrande and other transition pioneers—transitioneers. The genesis of the movement took place a few years earlier when Hopkins was teaching courses on permaculture at a college in Kinsale, Ireland. His courses covered projects on organic food production, natural building, woodlot management, nutrition, and other topics consistent with the permaculture design system, which I'll talk about in more detail later. Hopkins was also inspired by the work of Colin Campbell, the founder of the Association for the Study of Peak Oil, and Richard Heinberg, author of numerous books on the topic of peak oil. After realizing the profound implications that energy descent will have on virtually every community in the world, Hopkins made peak oil a central theme in his permaculture curriculum.

In September 2000, Hopkins and his students in Kinsale created the first substantive transition project that eventually came to be known as an Energy Descent Action Plan (EDAP). The objective of the project was to come up with a rigorous plan for the community of Kinsale with year by year steps to be taken. The goals were to help the community adapt to conditions of scarce and expensive oil, to reduce carbon emissions, and to become more self-reliant. After a draft of the plan was completed it was officially adopted by local authorities and was given its name, *Kinsale 2021—An Energy Descent Action Plan*. The Kinsale project became a template from which other EDAP projects have been developed.

A few years later, Hopkins relocated to the town of Totnes in Devon, UK, to work on a graduate degree. Inspired by the success of the Kinsale project, Hopkins met with Naresh Giangrande and others to develop a transition model for local citizen action that would support similar community-based projects. Based on this model, Hopkins, Giangrande, and other transitioneers piloted the first transition initiative, called "Transition Town Totnes." Drawing from his experiences, Hopkins gathered his notes and published them in a book, *The Transition Handbook*, in 2008. Since the first initiative at Totnes, the movement has gained tremendous momentum. As of this writing there are 360 official transition initiatives located in thirty-four countries, though all of these are still largely in the planning stage of development.

Transition initiatives begin with a core group of people who set out to raise awareness about the need to make the shift to a post carbon community and self-reliance. Most of the transition initiatives have adopted the model developed by Hopkins and modified them to fit the specific circumstances of their local communities. The model is basically a set of suggested guidelines for raising awareness about the issues surrounding peak oil and climate change, community organizing, training and reskilling, democratic participation, forging ties with local government agencies, and planning. But the most substantive part of the model is the EDAP. Up to now, most of the specific projects have centered on permaculture-based food production, EDAPs and renewable energy, and other projects that lead to sustainable and socially equitable outcomes. The transition model is still evolving and the initiatives are in many ways still in an experimental phase.

By their own admission, Hopkins and the other transitioneers acknowledge that the model needs to be developed further with a more comprehensive vision of economic systems and how they function. Up to now it has remained an open-ended question as to what kind of business models and economic institutions need to be developed to implement the EDAP and complete the transition to a whole, stable, resilient, post carbon local economy. I will be trying to fill some of these gaps in this model in the subsequent chapters of this book. For now, though, before examining the details of the transition model, it might be helpful to explore its philosophical foundations to get a sense of where it has come from. This will also help us take further steps in institutional development in ways that are consistent with the transition model.

Transition Philosophy

As it is Hopkins's field of study, the principles of permaculture lie at the foundation of his transition model. Like transition, permaculture is an evolving concept. The word "permaculture" is a contraction of the words "permanent" and "agriculture," which is another way of saying sustainable or ecologically permanent food production. Although permanent agriculture has been practiced for centuries in Japan and elsewhere, it has come to be noticed in the West more recently as an alternative to chemical-based industrial agriculture. In the last few decades, the Western view of permaculture has evolved beyond food production to include other aspects of human economic life.

In this broader sense, Hopkins defines permaculture as, "... a design system for sustainable human settlements . . . a template with which we can successfully assemble its various components—social, economic, cultural, and technical." Taking the idea a bit further, author and permaculturist Graham Bell defines it as:

> . . . the conscious design and maintenance of agriculturally productive systems which have the diversity, stability, and resilience of natural ecosystems. It is the harmonious integration of the landscape with people providing food, energy, shelter, and other material and non-material needs in a sustainable way.[9]

In this sense the basic goal of permaculture is to feed, shelter, and clothe people in ways that do not cause systematic ecological damage and therefore can be sustained indefinitely.

This notion of permaculture as an ecological design system extending beyond agriculture was founded by Bill Mollison and David Holmgren in the early 1970s. Their collaboration began when Holmgren was a young student at the College of Advanced Education in Hobart, Tasmania, and Mollison was teaching science at the University of Tasmania. Their effort to create a kind of living economy based on a set of core ethics and principles was truly a product of their time.

The decade of the 1970s was marked by profound developments in intellectual creativity and scientific discovery, particularly in the areas of system science and ecology. An environmental consciousness was emerging and people in general were becoming more aware of the impact their way of life was having on the planet. The first Earth Day was sanctioned by the United Nations in 1970 and a broad interest in ecological science was gaining rapidly. Of particular interest was the rate at which resources were being depleted, specifically oil. Oil production in the US peaked in 1970 and by the fall of 1973 a global oil shortage shocked the world. The shortage was an extension of the Yom Kippur war between Israel and a coalition of Arab states. As part of their war effort, the Arab states orchestrated a general oil embargo against the West in retaliation for the United States' military support of Israel. The embargo caused oil prices to take an unprecedented leap, prompting a deep global recession and rapid price inflation. The oil shock and the economic turbulence that followed revealed to scholars and scientists just how vulnerable our oil-dependent economies had become. Many were questioning the wisdom of our dependence on oil and sought out conservation measures and energy alternatives. Others questioned the materialistic lifestyles in the West and the culture of consumerism.

Aside from our oil dependence, many other things were being called into question at that time. Large numbers of people mobilized and demanded changes in gender roles, race relations, and how we interact with nature. A sense of change was in the air and the conditions were ripe for a significant paradigm shift in science. Social, political, and ecological complexities demanded a new way of thinking—a shift from what was considered an outdated and reductionistic framework of analysis to one more holistic and interdisciplinary.

More and more scientists attempted to break out of the confines of their disciplines that had become narrow and overspecialized. Overspecialization led to firewalls being placed around discourse, which thwarted communication among disciplines and created a kind of academic tribalism. Some of the more forward-thinking intellectuals were breaking away from this narrowness and embraced a broader purview that crossed traditional disciplinary boundaries.

One of the most significant contributions to systems thinking in science was by the Russian chemist Ilya Prigogine. Prigogine won the Nobel Prize in 1977 for his work on "dissipative structures," which asserts that science is nondeterminant. Scientific theories, when expressed in classic mathematical formulas, appear very pristine, tight, and deterministic. But the discoveries made in physics in the 20th and 21st centuries led physicists to draw very different conclusions. The physical universe is rather made of up of complex, chaotic, fuzzy systems that bleed into one another to the point that there can be no fundamental distinctions between the so-called hard sciences of physics and chemistry and the so-called soft sciences of human behavior. Prigogine's work helped form a bridge between what were previously isolated domains of inquiry. As it became more interdisciplinary, scientific inquiry became more inclusive and systems thinking began to replace reductionism. This naturally led to a more ecological way of seeing things.

By 1976 E. F. Schumacher published his collection of essays in a book titled *Small Is Beautiful: Economics as if People Mattered* which included an essay on "Buddhist Economics." Schumacher had a spiritual twist to his work and was also concerned about the spiritual well-being of humanity. Like so many others at this time, Schumacher was questioning orthodox science. He focused particularly on economic theory and its attempt to emulate the same deterministic models in the physical sciences that were being overturned by Prigogine. I'll have much more to say about Schumacher in a later chapter. However, it is important to note here that Schumacher was deeply concerned about our near-total dependency on oil and the foretelling of economic troubles the energy shocks of the 1970s presented. Four decades before Transition Town Totnes, Schumacher urged that economies move toward decentralization and re-localization in light of inevitable energy scarcity.

The genius of these intellectual pioneers of the 1970s gave us a new systems paradigm for understanding our place in the broader scheme of things. The distinctions separating physics from chemistry, chemistry from biology, biology from human behavior, human behavior from social systems, and social systems from ecosystems were fading away. It was in this rich intellectual environment that Holmgren and Mollison developed their principles of permaculture. In other words, this new systems and interdisciplinary framework made it possible to begin a rigorous study of food production and its relation to social and ecological systems.

Though Mollison is considered the grandfather of Western permaculture, it is the principles of permaculture developed by Holmgren in his 2003 book, *Permaculture: Principles and Pathways Beyond Sustainability* that were eventually adopted by the transitioneers. At the core of Holmgren's permaculture design system are three ethical principles: (1) meeting the needs of people, (2) doing so in a fair and equitable way, (3) and doing so in such a way as to not systematically damage our natural environment.

Extending from this ethical core are permaculture design principles that are conceptual tools or guidelines for directing people's work of meeting their needs in ways that do not result in environmental damage. Holmgren emphasizes that this work could be more consistent with permaculture design if the people doing the work are careful and observant about how they interact with their natural environment. Permaculture involves developing a process of learning through continuous interaction with nature and carefully observing the ecological impact of our work. It also maintains a principle of yield or bounty that results from our work. It is with the fruits of our labor that we meet our needs, but enjoying the quantities of stuff we produce should be balanced with a spirit of healthy ecology.

The work of meeting our needs cannot be done without harnessing energy. So another principle of permaculture is to be attentive regarding what kind of energy we use. Burning fossil fuels at a rate faster than the earth can reabsorb their carbon effluents violates the core ethic of avoiding systematic ecological damage. As we break away from our dependence on fossil fuels, we will be compelled to capture and harness renewable energy flows from the sun, wind, biomass, water runoff, and other sources that potentially do not harm ecosystems. Energy descent will force us to rely more on renewable energy, so valuing and vigilantly taking care of our renewable energy sources will be crucial. Though I should caution that no matter how vigilant we are in finding new and ecologically healthy energy sources, there is no hope of replicating the scale with which we torch fossil fuels. This points us in the direction of fundamental change in our economic institutions that are currently deeply dependent on fossil energy.

Another important principle in Holmgren's model of permaculture is to maintain a good structure of tight feedback loops. Feedback loops are an important part of learning how to sustainably regulate our interaction with natural systems. Positive feedback involves reinforcing whatever it is that we are doing now so that we can do it even more in the next cycle. For example, if we capture and store energy in a way that enhances our ability to capture and store more energy, then the amount of energy available for use will increase exponentially. Positive feedback doesn't necessarily mean something good, however; it just means a process of self-reinforcement. If this exponential growth in energy leads to exponential growth in economic production, eventually our ability to harness renewable energy will be overwhelmed, causing the self-reinforcing process to break down.

Such a breakdown is a negative feedback process. When a PA system gets caught in a positive feedback loop and creates an awful screeching sound, eventually either the speakers will blow out or someone turns the power switch off. Either case is negative feedback. Just as positive feedback does not necessarily imply something good, negative feedback is not necessarily bad. Positive and negative feedback are key aspects of systems thinking because wherever there is interaction among elements within a system, there is feedback in both directions. Since interaction is everywhere, feedback processes are also everywhere. Systems can become self-regulating so that there is a balance maintained between positive and negative loops. An example of self-regulation in nature would be the dynamic balance between predator and prey that keeps population numbers relatively constant. A prey species growing in numbers through reproduction is a positive feedback process—the more animals there are, the more offspring they produce—and their numbers grow exponentially. But this growth will attract more of its natural predators as a food source. As the predators eat the prey, this acts as a negative feedback process that keeps the exponential growth in check.

However, I also caution here against using such feedback processes in nature as a model for human social and economic systems. As our economies are programmed to grow exponentially, we are drifting toward a condition in which resource depletion will force the growth to stop. Resource depletion and the crises that will follow would be negative feedback like the blowing out the PA speakers, and the conditions for human life would be rendered pathological. The advantage we humans have, however, is our conscious ability to see these things clearly and alter our behavior accordingly. Such proactive behavior is a healthier condition of negative feedback like throwing the switch before the PA speakers blow.

Forming tighter feedback loops can be important for steady state economic systems in other ways through communication—another outstanding human capability. A key aspect of economic re-localization is proximity. How quickly we can adapt to changing circumstances will depend on the effectiveness of communication within them. That will also depend on how close producers and consumers are to each other. If consumers say, "We want the producers to stop putting chemicals in our food supply" then it is much easier for those producing the food to respond to that feedback proactively if they are located in the same community rather than 5,000 miles away. When people are closer to the origin of the things they consume, they tend to be more conscious and aware of how those things are produced. Those who produce those things, in turn, become more aware of how they are consumed and the result of that consumption: Is it making them healthier, happier, sick, depressed? This will prove to be an enormous advantage with smaller scale economic re-localization.

Another key principle for permaculture and sustainable living is to refrain from allowing human-produced waste or substances taken from the crust of the earth to accumulate in the ecosphere. Shorter feedback loops made by tightening the proximity between production and consumption can help people monitor the sources of pollution and the accumulation of carbon, toxins, plastic, and so on.

It is important to keep in mind that permaculture is a design system. One of the core design principles is to develop production systems for food and other goods such that they merge naturally with patterns in the surrounding habitat. The natural habitat can be seen as a kind of structure or frame around which the details of the production system can be positioned. Producing food, harvesting energy, and making other value-added products from vegetation can be done naturally within the confines of the patterns that occur in the surrounding habitat. This is integral to Holmgren's principle of interaction and observation. As we observe our surrounding habitat carefully, we can discern natural patterns in how things grow and what kinds of plants and animals thrive and what kinds don't. If someone plants something that starts behaving like an invasive species, the person can observe and learn from the experience, then try something else. We can keep trying alternatives until we find something that meshes well with the natural environment. Similarly if we try certain manufacturing processes that burn more energy than what can be renewed, we can learn from that as well and change our processes. Structuring human activity so that our interaction is synchronized with these patterns can go a long way toward achieving true sustainability. Of course this would be a radical change from the way most food and other goods are produced now. This, again, points in the direction of institutional change.

As we continue the work of evolving our systems of production, we can also see how each element of the system interacts with the other elements. A basic tenet of systems science is that the interaction among the parts of a system is as important as the parts themselves. From that interaction, new things emerge that are distinct and unpredictable. Holmgren contends that integrating as many elements as possible to create richer and more complex systems is a healthier design system than segregation and exclusion. This applies to social systems as much as ecological systems. Including a rich mixture of diverse elements in both social and natural systems can lead to more diversity, and diversity is key to resilience.

The opposite of diversity is conformity or singularity such that all the elements in the system are the same. Monoculture agriculture works this way in which only one type of food is produced over vast acreage of farmland. This is a poor design for food production as the farmland becomes more vulnerable to swarms or disease. Large scale agricultural systems in which everything is exactly the same will attract armies of bugs and pests of equal proportion. Diversity is nature's way of providing stability by allowing various species of fauna and flora to adapt and co-evolve into complex structures. Again, this principle is applied to human communities as much as it is to systems in nature. A complex balance of uniqueness and differences can enliven human cultures, which enhances people's ability to adapt to changes.

Diversity is closely related to another permaculture concept known as "modularity." Modules, distinct and somewhat autonomous units within a system, also stand as parts of the system. The principle of integration or inclusion is important, but it is also important to assure that each component can rise and fall on its own without taking down the entire system. Although mutual support among the modules is also important, it would be a poor design if the system died once any one particular component lost its functionality.

The points of contact between components of a system as they interact are also important. On this Holmgren identifies another principle emphasizing the use and value of the edges and margins of a system. Topsoil, for example, is an edge between the crust of the earth and living beings, our skin is an edge between our bodies and our surroundings, and an estuary is an edge between a river and the sea into which it flows. The point of contact between elements in a system is where interaction takes place and if the elements are neglected or allowed to become unhealthy, then the health of the overall system will be compromised. In economic systems the edges or points of interaction between an individual person and the overall system are the economic institutions. These basically structure the work that we do, how we do it, the reasons why we do it, and so on. Conscious care in institutional design will be of paramount importance for true transition efforts.

It is also important to be conscious of the scale and pace of this structured work. Holmgren contends that small and slow solutions to problems in the design system are better than big and fast ones. Keeping our production systems limited to the smallest possible scale will also minimize the impact we have on our natural systems. Keeping the pace as slow as possible will serve the purpose of sustainability by not using resources at a pace beyond the natural pace of recycling or regeneration capacities. Again, though, this principle has to be balanced with the ethical principle of providing for the needs of the members of the community.

Finally, there is the principle of change. One of the central tenets of systems thinking is that all systems are constantly in a state of flux and transformation. It is an immutable scientific fact that everything is in both a state of being and becoming. Our ability to survive the end of the Oil Age will depend on how well we thoughtfully respond to inevitable changes in both our natural systems and in our own cultures. Cultural lag—the socially rigid resistance to change—leads to stagnation and decline, whereas cultural adaptability leads to vibrancy, resilience, and life in our communities. Rigid resistance is like a brittle twig that will snap in the winds of change and flexibility is suppleness and adaptability. One of the key challenges to our work of building new economic institutions is to create them in such a way that they remain true to their mission, but to avoid excessive rigidity so that they cannot adapt to changing circumstances.

In summary, the set of principles for permaculture design systems emerged in the 1970s as a part of a paradigm shift in science. The shift was a movement from the antiquated, reductionistic, and mechanistic model for scientific inquiry to a more holistic systems approach. Just as traditional methods of scientific inquiry were being called into question, so were the methods of industrial agriculture. The old model of seeing everything reduced down to isolated billiard balls swirling around in empty space was replaced by a vision of complex systems with properties derived from dynamic interaction. In the new vision every phenomenon—inorganic, biological, mental, social, or ecological—is part of a dynamic system. Mollison and Holmgren and others expanded on this framework to establish a design system mostly centered on food production, structured around the core ethics of meeting the basic needs of people, but doing it with a commitment to justice and ecological sustainability. The transition movement added another facet to this by applying these principles specifically to the problems of peak oil, climate change, and global economic instability.

The Ingredients of the Transition Model

Though the transition movement is rooted in the core ethics and principles of permaculture, Hopkins and the other transitioneers are careful to avoid dogmatism. They do not issue directives on how people should live, rather they emphasize that the work in the community should be allowed to flow where it wants to flow naturally. Rather than giving directives for prescribed behavior, their transition model offers guidelines which the members of the community may or may not follow. They refer to their set of guidelines as "ingredients" because it is more like a recipe from which each community can choose to use all or part, depending on their circumstances or culture. Communities are encouraged to improve on the guidelines if they can by inventing new ingredients. Nonetheless, the guidelines have an overarching purpose: to help communities reduce carbon emissions and prepare for a future community without oil.

Among the first ingredients in the model is to form a steering group to get started and to identify how the sphere of the transition work will be defined. In other words, the group will determine whether the sphere will be limited to neighborhood, community, town or city, or given some other definition. Once this is defined, the transitioneers recommend forming subgroups straight away. Each subgroup will draw on the talent and experience of individuals within the community so as to focus on particular aspects of the transition work. The main idea of the subgroups is to home in on a specific project, while at the same time staying focused on the mission of the movement overall. In the transition movement in my hometown, Portland, Oregon, known as "Transition PDX," we have seven subgroups: food, neighborhoods, planning and policy, community preparedness, heart and soul, publicity and communication, and local economy and business. They are in continuous communication with the larger group through a local listserv. Typically the subgroups hold regular meetings and then report to the larger, umbrella group in an open space meeting environment.

The open space meetings bring interested members from the community together in a large group to discuss the various transition topics with no predetermined agenda, timetable, or designated coordinators. The idea behind open space is to devolve power to the community by allowing a free association of ideas where everyone who wants to gets a chance to make a contribution. One aim of this approach is to foster open-mindedness and genuine learning. Another is to create an environment in which people are free to gather and explore issues related to the initiative without fear of judgment, or without feeling like they are being trapped into serving someone else's ulterior motives. All participants are free to contribute to the direction of the collective conversation. The hope is that through this open process in which all the themes and visions have been explored, a consensus on what is most important will naturally bubble up the surface. By creating a more democratic meeting structure they also hope that this will serve the core permaculture ethic of fairness.

The transitioneers are committed to a grassroots, democratic approach to governing their initiatives. To ensure this, they suggest that once the subgroups are formed and are able to carry out their work, the steering group should be dismantled and a representative of each of the subgroups take over positions of the initiative's governance as a kind of initiative council. The assumption is that steering committees can often get entrenched in the particular agenda of individuals with powerful or persuasive personalities. Such an agenda may or may not be appropriate for the project. If an organization is formed around the personal agendas of the steering committee individuals, the organization itself can take on a life of its own with an instinct for survival and self-protection at all costs. Such entrenchment can thwart progress toward the goals of transition.

Dissolving the steering committee, however, can be tricky. In order to be successful, a movement must remain true to its mission and values. A steering committee or board of directors' primary responsibility is to make sure that the movement or organization stays on the rails and does not get sidetracked or co-opted by some other agenda. This has to be balanced with the flexibility to adapt to changing circumstances. The movement will not be able to succeed if it is subject to domination by a small group of self-interested steering committee members, nor will it succeed if it becomes wishy-washy and loses its focus to factionalized subgroups that themselves become silos of self-interest. Effective governance requires an oversight body whose primary responsibility is to resolve conflicts and make sure that the subgroups in the initiative are doing the real work of transition. Perhaps this oversight body can be made democratically accountable both through some kind of electoral process by people within the movement as an electorate and by a kind of constitutionality of the initiative such as a mission statement, statement of guiding principles, or articles of incorporation.

However the initiative is established, once a structure of governance is in place and the subgroups are formed, the next ingredient is to start raising public awareness. Getting the word out about the issues surrounding climate change, peak oil, and community-based solutions, as well as the movement and its goals is labor-intensive. Raising awareness has been the major part of what participants have been occupied with up to now. The work involves community education courses, showing films, hosting neighborhood talks and discussions, presentations by experts, articles or programs in local media, and community forums. Here there is much emphasis on maintaining as much positive energy as possible and at the same time addressing issues specific to the community. Awareness-raising is calling attention to what people stand to gain by being proactive and what they stand to lose by remaining apathetic.

Talk of the potential dangers and crises that emerge from peak oil and climate change should not be dismissed as mere doom talk, yet emphasis on the possibilities for positive change should not be ignored either. Both are key parts of raising awareness: awareness of the dangers and of the possibilities.

As the people are more informed and become more involved, additional groundwork needs to be done to establish momentum. One way to do this is to form ties with like-minded organizations that have already been working on these issues. Nearly every community has at least some active citizens or organizations that have been working on projects that would be aligned with transition goals such as renewable energy, permaculture, urban farming, or economic re-localization. This work should be honored and acknowledged. Also to be honored are the older people who lived during times when households were less reliant on the global economy and more self-reliant. It is cautioned, however, not to be suggesting reverting back to older times. Stitching together a raft with these groups and people gives the transition power in numbers and serves as a catalyst for propelling the initiative into its next phase that transitioneers call "The Great Unleashing."

The original transitioneers launched their first unleashing in Transition Town Totnes in 2006 after ten months of careful planning, talks, films, and speaking engagements with an EDAP in place. This stage of the transition initiative is primarily symbolic and inspirational. The idea is to bring people up to speed by launching a memorable celebration that will also serve as the transition's kickoff. Whatever suits the culture of the community or town— food, local musicians, artwork, storytelling—the idea is to begin the work of transition in a festive spirit and to celebrate the resourcefulness and creativity of the members of the community. Creating positive energy in this way will help break down the inevitable barriers or roadblocks to change—either psychological, personal, or institutional. Beyond the mood of celebration and adventure, the community will still want know or see something functional in the unleashing. So part of this kickoff event should be a discussion of the EDAP and express practical indicators of how the project could be successful.

Helping communities devise practical ways to adapt to a continual decline in available energy is the core mission of the transition movement. When people in the movement refer to energy descent, they need not only mean the troubles surrounding the declining availability of oil. They can also strike a positive chord by pointing to communities that are successfully re-localizing their economies and becoming more self-reliant. Using the EDAP to detail a series of practical steps in this process could prove to be a good way to launch a local initiative.

Within the global transition movement numerous EDAPs have been unleashed. Each plan is different and specific to the community's initiative, but certain key elements have emerged as universal and essential. One such element is to gather baseline data on existing energy use and sources. If a transition initiative is to have an energy descent plan, it needs to have some idea about how much energy is being consumed and from what source that energy is generated. Is the electricity provided by public or private utilities? Are these utilities powered by coal, hydroelectric, wind, solar, or a combination? To what extent do people rely on public transportation or cars? What are the typical commuter miles? What are the typical food miles? How much arable topsoil is available both within and around the community? These and other energy-related questions will give the transition community a sense of what work needs to be included in the plan. Most local government agencies have already compiled some of this data as well as their own long-range strategic and economic development plans. If the initiative integrates its EDAP with existing plans local government agencies have for infrastructure and economic development, the plan will be more effective. It will also be important to know what kind of industries the government agencies plan to support or subsidize, as well as how they are preparing for population growth, land use, traffic management, and other aspects of municipal or local government.

It is essential that the transition initiative forge ties with these and other local government agencies. Schools boards, transportation authorities, planning commissions, city councils, and other government bodies can play a vital role in genuine transition initiatives. However, making this connection can be difficult because government agencies are often cumbersome, bureaucratic, and are most often embedded within business-as-usual structures of power. Such embeddedness often creates a climate of distrust. Nonetheless, this has to be overcome and can be overcome with a vocal and active citizenry, and by doing away with the "us versus them" attitude toward local authorities. Most important is for the transitioneers to cultivate a positive working relationship with agencies at all levels. The transitioneers caution, however, that local governments are most helpful if they play a supporting role in the initiative than directing it.

In their supporting role local governments can generate useful data, set ordinances and growth boundaries, develop better public transportation systems and recycling systems, waste management, water conservation, funding, and so on. Again, citing Portland as an example, the city has become renowned for sound planning as it has for decades been proactive in making the city as fossil fuel-independent as possible. Portland's infrastructure includes a modern light rail public transportation system, inner-city trolleys, and bicycle boulevards. In 2011, the city unveiled a draft plan to, among other things, reduce carbon emissions to 50 percent below 1990 levels and to establish complete communities for 90 percent of the city's inhabitants by making all the necessary amenities available within a twenty minute walk radius by 2035. Whether they accomplish these goals remains to be seen. The plan is the product of collaboration between the city's bureau of planning and a host of local schools, universities, businesses, neighborhood associations, and a long list of other interested groups. This is not specifically a result of the transition initiative, but the local government's planning process fits the transition model. We will explore the Portland Plan in more detail in a later chapter. Minimally, however, local government agencies can be a good source of data and information that will be key to composing a workable EDAP.

Once all the information is gathered with a solid understanding of the energy situation in the community, and with local government on board, a long-term vision for energy descent can be created. The EDAP can start with broad strokes and project indicators for how much CO_2 emissions and fossil fuel consumption will be reduced over a specified timeline. The plan should detail how much renewable energy could be realistically harnessed within that time. It should try to extrapolate as much as possible the impact and changes in people's lives these energy reductions will cause—positive and negative. From there the plan can be fleshed out with more details segmented into areas of focus including: food, energy, transportation, waste, planning, communication and publicity, health, education, economics, local government, and water.

The EDAP and specific hands-on transition projects are the most substantive aspects of the initiative. Raising awareness and planning are important, but unless the transitioneers can show the community something palpable, it will be difficult for the initiative to gain momentum. Small noncontroversial projects can provide a visual reminder that real, material changes leading to self-reliance and resilience are possible. This can inspire members of the community to get more deeply involved in the work. Presenting visible progress or at least the potential for visible progress in achieving the transition goals will go far in getting community participation. Other visible projects are more technical and also require skills and training. Developing these new life skills and technical skill sets that augment the work of rebuilding a community for a future without oil is what transitioneers call "The Great Reskilling."

The requirements for reskilling are enormous and communities will have to use every resource available. As I mentioned above, one of the ingredients of transition is honoring the knowledge of our community's elders, particularly ones who lived before planned obsolescence and economic globalization. The word "reskilling" is apt because some of these skills are ones that people used to have, but were lost as one community after another was drawn into the vortex of the global economy. Many of the older generations learned how to can food, make clothes, and repair things, skills that are almost completely unknown to the deskilled younger generations. Older people can also help the young understand the meaning of real communities in which people lived: towns or neighborhoods where they knew the local farmers and shopkeepers. These sensibilities have also been lost on many of the younger generations who were raised consuming products designed not to last, but to be thrown away and replaced as quickly as possible—many of them imported. In the process of making our local communities more resilient and self-reliant, younger generations will need to learn many new practical things as we all begin our energy descent.

Energy descent will dissolve the current structure based on the global division of labor. This implies that ordinary people will have to learn how to produce, repair, and maintain things themselves. To this end, transition communities have been holding workshops, courses, seminars, and teach-ins on a range of related topics. Though incomplete and evolving, here is a list of these skills suggested by Hopkins and others:

> Cob building construction and design, tool sharpening and maintenance, food preservation, energy auditing, solar panel installation, water mills and generators, making biofuels, domestic energy efficiency and conservation, house retrofitting, composting, conflict resolution, community organizing and leadership, communication, local currencies, clay plastering, making earth paints, making herbal tinctures, household finance, scything, straw bale building, knitting, weaving, sewing, dyeing, tree pruning, horticulture, stove making, oil vulnerability auditing, music, time management, indoor and outdoor gardening, bread making, biodynamics, nutrition, bricklaying, appliance

repair, welding and metalsmithing, carpentry, cabinet and furniture making, brewing, banking, chartering new business models, merchant card services, retailing, light handicrafts, and the list goes on.

The more these skills are mastered by people in the community, the more diverse the economic activity in the community becomes. As a core principle of permaculture diversity is centrally important for a transition to economic re-localization. More often than not, the word "diversity" is used in the context of cultural diversity and fairness in the structure of opportunity. But here the term has a much broader socio-ecological meaning. Diversity is essential for any system's ability to adapt to changing circumstances and environment. Nowhere is this more important than in food production.

Because of its obvious importance and connection to permaculture, food production is a major part of transition initiatives. Growing food will become more difficult as climates change and basic resources like energy and fresh water become harder to get and more costly. This will compel communities to secure their food supply with sound practices, including diversification. Farming is unlike many other businesses because it is so utterly dependent on weather, soil, and climate conditions. As these conditions change with global warming, the reliability of food production is jeopardized and crop failures become more likely. Crop failures have always been a frustrating but unavoidable problem, and this will get worse as the condition of the soil is damaged and climate becomes increasingly erratic.

The best defense against crop failure is to diversify. Carol Deppe in *The Resilient Gardener: Food Production and Self-Reliance in Uncertain Times* (2010) suggests that we can hedge against uncertainty by growing more than one kind of crop at different times of the year. For example, instead of relying exclusively on strawberries for a spring fruit, we should also grow cherries. Not only will they be affected by changes in climate in different ways as one is a bush and the other a tree, but the process of harvesting the fruits requires different activity. One is harvested by labor performed close to the ground and the other by climbing ladders. Growing both kinds of fruit, will add rich texture to the landscape and will create new interactions with different types of pollinating insects. Diversity could be extended by, say, adding apple trees as well. Deppe makes the very practical argument that if the farmer fell off a ladder in the spring and was unable to harvest the cherries due to injuries, the injuries would hopefully be healed by the fall when the apples were ready to harvest.

Careful consideration needs to be paid to all the factors of food production such as the availability of labor at certain times of the year. Are people preoccupied with harvesting one kind of labor-intensive food product and therefore unable to spare the time to harvest the other? Will the other crop then go to waste? Are some types of food more vulnerable to extended drought periods or periods of heavy rainfall? Diversity and resilience go together because with diversity, one crop could fail while the other is produced and people would still have food to eat.

Diversity can extend beyond primary food production as well. Secondary production such as drying, processing, canning, from the food is important for preserving. But there are other occupations that go beyond food that perhaps people should also learn, such as furniture restoration, repair work, salvaging and recycling, or handicrafts. The global division of labor in our established economic system has led to extreme levels of specialization in which entire economies are completely dependent on a single industry like oil refining or high tech. If those industries become destabilized for whatever reason, the entire economic system fails. Simply put, diversity is not putting all your eggs in one basket, as the saying goes. It enhances adaptability which is inseparable from resilience.

Helena Norberg-Hodge, director of The International Society for Ecology & Culture (ISEC) makes a compelling case for the inseparability of food resilience, ecological diversity, and cultural diversity. A pioneer of re-localization as an antidote to the destructive and homogenizing forces of globalization, she argues that rebuilding local knowledge specifically derived from learning how to use local resources and developing context-specific skills is vital to healthy communities. Like monoculture agriculture, globalized Western style consumerism supplants the direct relationships people have with their natural surroundings and tight-knit communities, which eventually wears away at the cultural and ecological fabric. Local knowledge gets lost in time. Norberg-Hodge writes:

> Revitalising local knowledge is essential to localisation. Without retreating into cultural or economic isolationism we can nourish the traditions of our own region. A true appreciation of cultural diversity means neither imposing our own culture on others, nor packaging, exploiting, and commercializing exotic cultures for our own consumption. . . . We are transporting across whole continents a vast

range of products, from milk to apples to furniture, which could just as easily be produced in their place of destination. . . . Local agricultural knowledge is essential to knowing which crops are adapted to the local environmental conditions. Producing food with this sort of intimate understanding of the land and climate enables farmers to work in a much more sustainable and efficient way. It is in robust, local-scale economies that we find genuinely "free" markets; free of the corporate manipulation, hidden subsidies, waste, and immense promotional costs that characterize today's global market.[10]

As people in a self-defined local community are able to do multiple tasks and produce a multitude of things to meet their needs, the structure of the community becomes more complex. Transitions are about re-weaving the web of relationships among people and organizations as they perform functions on different levels. If the entire community revolves, say, around corn production, and depends on others on the outside for everything else, then there is little complexity—there are the land owners and farm workers and little else.

As the web of complexity is woven, feedback structures emerge within the connections. A more localized and diversified economic community will have shorter distances between producer and consumer and the communication of what is working or not working will be more direct and effective. In a globalized corporate system, communication is dysfunctional and impersonal. Anyone who has been put on hold by an automated answering system can attest to this. The work, including customer service, is performed halfway around the world and it is virtually impossible to make necessary adaptations because of the distance, bureaucracy and weakened connections between people and between people and organizations. Tighter connections among producers and consumers naturally lead to more accountability and this can serve the need for fairness as well.

As a re-localized economy evolves into an actual system it will give rise to financial institutions, manufacturing, retailers, cooperatives, and so on. Each of these elements will be tied to every other element in the system with tighter and well-observed feedback mechanisms. Every individual and every business model will be contributing directly to the evolution of the new system. We will explore the process of building economic institutions later, but it is important for the transition initiative to put as much care as possible into structuring these institutions according to its core ethics. In short, transition needs a values-based economic paradigm.

There is no blueprint for what a transition initiative should look like. The process amounts mostly to forging ahead into the unknown. Doing the work of building a community without oil is doing the work of the future, the unknowable. People like to talk about being on the "cutting edge" when it comes to technology, but being on the front lines of a social movement never feels so sharp and decisive. It is more like being on a ragged edge because the process is messy, bruising, and uncertain. Granted, it is important to try to have as much guidance as we can to give us direction, but since we are trying to rebuild our communities into something that does not yet exist, to some extent we simply have to make up the initiatives as we go.

Real transformation comes from doing things. We cannot change the world by thinking about it; we change the world by actually getting out there and doing things in a different way. And the more we get out there and do things in a different way, the more it will change our habits of thought. And when we break out of old habits of thought, we begin to see everything in a new light.

Transition Projects

People are doing things and they are effecting change. In the US there are nearly a hundred transition initiatives underway at different stages of development. Most are in the early stages and concentrating on raising awareness and forming working groups. Nearly all the transition initiatives involve lectures, discussions, film screenings, and book readings related to transition work. But there are also many reskilling events going on everywhere.

Many initiatives are offering courses and workshops on permaculture. But some have gone beyond the permaculture basics to teach people things like how to build "herb spiral" gardens in which natural gardens are laid out in beautifully aesthetic nautilus shapes. Others are organizing field days programs to instruct food growers on how to break away from their dependence on petrochemicals by using natural pest management systems with insects, birds, animals, and plants as well as human-made technology to keep down the population of destructive pests.

One initiative organized a Neighborhood Gardening Guild. Plant guilds have proven to be an effective strategy for building sustainable food landscapes. A plant guild is a collection of mutually beneficial plants that live together symbiotically in such a way that a broader plant community can thrive. The idea behind the Gardening Guild is to create a social network that mirrors the symbiotic interaction of sustainable landscapes and natural vegetation. Just as a plant community thrives as a result of the mutually beneficial interaction, people in a community who work together by growing and sharing food can also thrive.

Another initiative has formed a community-based garden wheel. A garden wheel is a barter community in which people exchange work directly for locally grown food or seeds or something made from locally grown food. The principal notion is to create more local economic transactions to reduce food miles and therefore reduce fossil fuel use. Some projects focus on food equity as well such as creating sharing gardens which are like community gardens, but the surplus food is donated to the local food bank. Small plots of land have been converted into learning gardens where members of the local community can learn more about ecologic stewardship, poly-culture garden beds, and have herb and vegetable stands where the learning garden can sell its produce to the general public and use the proceeds to develop and expand their projects. In other locations people are learning to plant and care for fruit trees, do appliance repair, operate aquaponics systems, stage "eat local" festivals, and in one initiative a local chef offers workshops on how to sharpen knives and basic cutting techniques for food preparation.

Not all transition initiatives are centered on food. Some transition activists are organizing reskilling workshops on a variety of do-it-yourself projects such as growing native plants, graywater and rainwater harvesting, cob and strawbale building design and construction, and a project on how to build a house out of salvaged wooden pallets. Some are hosting seminars on herbal medicine, on making fiber art from raw wool dyed with natural plant extracts, or how to make blankets out of repurposed fabrics—fabric that was originally produced for something that was subsequently dismantled and used again for something else. Some are hosting forums on health care and organizing for better legislation. And everywhere are initiatives centered on expanding solar power infrastructure, encouraging reduction in energy consumption by using more efficient electrical appliances and lighting, and lobbying local authorities to revamp transportation systems.

Some transition projects are initiating experiments in localizing money and other exchange systems as a way to solidify economic re-localization. There have been numerous experiments with paper currencies to be used only in a particular town or city. Some local currencies have been more successful than others. Local paper currencies are primarily used to facilitate local transactions, and to keep the money circulating locally. The process of implementing a paper currency is relatively uncomplicated and perfectly legal. Once the local currency is issued, usually through a local nonprofit, it can be used as a medium of exchange at local stores, barter clubs, or trade fairs. Community members can use their local currency to buy bread from a local bakery or pay a house painter or dentist for their services. The principal advantage, of course, is that money earned and spent in the local economy stays there, which will hopefully add vibrancy and resilience. Keeping local economic activity alive and resilient will be essential for sustaining people's livelihoods as the globalized division of labor becomes suitably obsolete.

Similarly, Local Employment and Trading System (LETS) have attempted to achieve the same goal of local economic vibrancy without the paper currency. LETS systems are electronic trading credit systems. A LETS system is also typically organized through a local nonprofit organization, which provides electronic "clearinghouse" services for its members. It is a system that is patterned after services that specialize in check-clearing. Standard banking clearinghouse services process check-clearing services for banks in which someone writes a check and sends it to another. In the transaction, the receiver's account is credited or added a monetary balance and the sender's account is debited or reduced. LETS carries out similar accounting services by clearing transactions directly through electronic means without paper currency, checks, or commercial banks. The idea is that the services are contained within the self-defined community to hold member merchants and consumers together in a community-based trading system.

Time banking is a system of exchange that uses units of time as a medium of exchange rather than money or currency. Time banks have been around since the early 1980s and the process begins with forming a time bank institution in a local community. As people open accounts in the time bank, they can perform work, usually some kind of service, for another member. But rather than receiving cash they earn time credit measured in hours. These credits are sometimes called "time dollars" and constitute a kind of nonmonetary "deposit" in the time bank. The more work performed in this way, the more time credit one accumulates in one's time bank account. That time credit can be redeemed to receive services.

All these projects have the potential to contribute to the transition movement's overall goal of meeting the needs of people and community while kicking the fossil fuel habit. The transition movement is still in its early stages and evolving. The work people are doing in these hundreds of initiatives worldwide is cementing a foundation upon which more work can be done. As the movement evolves, however, it will have to overcome certain challenges.

Challenges to the Transition Movement

As I presented earlier, the philosophical foundation for the transition movement is a holistic, systems approach to science and the principles of permaculture. A holistic and ecological framework would have to integrate social systems with ecosystems. A recurring theme in this book is the role that institutions play in structuring people's work within these social and ecological systems. Aside from working with local government agencies on developing EDAP and land use planning, there is very little mention of social institutions—particularly crafting new institutions—and very little on economic institutions in the transition literature and projects. Crafting new institutions that are consistent with the goals of transition will be an important extension of transition initiatives.

Understanding the nature of social institutions is indispensable if we are to understand how our economies function, and it's even more crucial for understanding social change. If there were ever a case for institutional change, it would be to create new rules and norms of social behavior in the local community so as to effectively prepare for a time when cheap oil is no longer available.

As they are averse to dogmatism, transitioneers are very keen to avoid prescribing rules or norms for behavior in their initiatives. Though it stands to reason that if established prescribed social behavior has brought us to the brink of social and ecological ruin, then those prescriptions will have to change. Moreover, if the established prescribed behavior is structured around fantasies such as a never-ending flow of cheap energy, then again, that will have to change.

It is true that people don't like to be told what to do or how to behave. But it is also true that they are constantly told what to do and how to behave by corporate media; and, sadly, so many in our society just slavishly follow. The parameters for behavior set by corporate media are subtle. The question to be raised here is not whether or not there are institutional forces shaping our behavior. The question is how people in their communities can gain democratic control over them. If people deny the presence of institutions and the empowerment that can be derived from proactively creating them, then the result would be dangerous myopia in which a large swath of the socio-ecological landscape will remain unattended and allowed to fall into ruin.

Let me put it another way. Without institutional change, business people will continue to operate out of conventional business models with very specifically prescribed corporate behavior directed at maximizing profits and accumulating ever-expanding amounts of wealth. This will render every ethical principle that underlies both permaculture and the transition movement impossible and irrelevant. The gap that has strangely separated economy and ecology will widen. As much as green or ecological economists claim otherwise, we are presently facing a conundrum: the institutional forces that drive businesses depend on the ecological forces that support life, but the ecological forces that support life are incompatible with the institutional forces that drive business.

This conundrum is a kind of dueling myopia. If the people who are committed to protecting our natural habitat ignore the institutional prescriptions of business behavior, then they will not be able to see how we have come to the brink of total ecological ruin. They will be unable to make any connection between their environmental goals and economic processes. On the other side, if business people carry on with business as usual and ignore the ecological threats to their own survival, then they will be unable to make any connection to a truly sustainable ecosystem. There have been many claims that the prescribed rules of corporate enterprise and the ethical principles underpinning sustainability are not in conflict. Most of these claims, I believe, are dubious and I will explore them in detail in a later chapter.

The discord between business as usual and our ability to deal with energy descent is real and it is with this discord that the transition ship runs aground. If we resist institutional change by clinging to business as usual, then we cannot protect ourselves from the psychological despair and social crises that energy descent and global warming will rain down on us. As we try to live with glaringly obvious contradictions like endless growth driven by corporate institutions and the reality of a finite planet with limited resources, this will tear away at us, psychologically sending us deeper into states of delusion.

The avoidance of institutional forces also leads to the "lifestyling" of our economic and ecological problems. Lifestyling our problems suggests that we all live in a society free from institutional control where the direction of our collective future is determined by the lifestyle choices that we as individuals make. The assumption is that as people make choices according to their values, the fate of society as a whole will be determined. This stands in denial of the power that large corporations and the corporate media have in defining the parameters within which those choices are made.

To ignore this would be remarkably naïve. For over a century corporations and other big money interests have been confident of their ability to mold choice-making behavior with massive investments in advertising. Spending on corporate advertising in the US is larger than the gross domestic product of most countries in the world, and businesses wouldn't be spending this kind of money if it weren't effective. Marketing specialists realized long ago that if the direction of society is going to be determined by choice-making behavior, then they are going to do everything within their power to influence those choices.

Deliberate lifestyle changes will be a part of the overall changes that have to be made, but these changes cannot be made in isolation from an institutional context. Institutions define the parameters of our habits of thought and action. By necessity, real transition work has to be done in conjunction with creating new institutions that will redefine those parameters. This means creating a whole set of new rules and norms for social enterprise and money.

If there were anything that would constitute a primal economic institution it would be money. Because money is the veritable lifeblood of any economic system it is closely guarded by established institutions. Nonetheless, a transition economy will need to forge ahead to create new localized monetary institutions. Up to now, most of what transitioneers have tried are various attempts to create local currencies, LETS systems, and time banks. Though on the surface these seem viable ways to re-localize economies, they've proven to do very little toward that goal. Here again the problem is institutional. Transition activists are on the right path as they talk about barter communities and local money because they recognize the simple fact that if you can control the money, you can control the economy. This is precisely why within our established system, command central is the large Wall Street commercial and investment banks and their associates at the Federal Reserve Bank.

But printing our own Monopoly cash or trading credits is not the solution. A major disadvantage of local currency is that it cannot work well as a financial instrument to be saved. The main reason for this is that once the money is saved in a bank, it automatically becomes subject to the institutional control of the Federal Reserve System. When that happens, it immediately loses its standing as a local currency. The Federal Reserve System also requires that a local currency cannot be used to transfer funds from one bank to another. Local currencies are not used to make bank deposits as these deposits, or reserves, are what the Federal Reserve uses for controlling the money supply. If a local currency were used to make deposits, it would be, in effect, posing as US dollars. And when a currency starts posing as a dollar, it would be more effective to surrender and actually be a dollar.

A central problem in the LETS system is that it is prone to instability. A key problem with any monetary system is determining how much money should be maintained in the system and how to get that money into the system in the first place. LETS systems do not have a regulated money supply and allow unlimited transactions. Without strict controls on the availability of credits in the system people could buy unlimited quantities of goods, running up unlimited balances and—like printing off too much paper currency—the system will very quickly become unstable and potentially collapse. In fact, most experiments in LETS systems have collapsed for this very reason.

Time banks also have their problems. One of the underlying assumptions of traditional time banking is that an hour of one person's labor is of equal value to an hour of any other person. This assumption is, of course, dubious as it would be hard to convince members that an hour of babysitting is of equal exchange value as an hour of an electrician or plumber or surgeon's services. Most time banks were originally created as a substitute for social services and have had that social service orientation: child care, elderly care, legal assistance, language lessons. These services can and do provide for the needs of people through public institutions. But here again, people in their communities need to understand the exact nature and function of these public institutions and support them if they are to continue.

At best, time banks, LETS, and local currencies can help people within an already self-defined community share, barter, and exchange system. But the limited and cumbersome nature of these alternatives will prevent them from becoming a part of a fully developed economic system unless there is a more serious commitment to real institutional change. I have some other ideas for creating new financial institutions that can serve the purpose of economic relocalization and will also explore this in a later chapter.

Conclusion: From Transition to Transformation

I prefer the word "transformation" to "transition." Transition implies a lateral shift from option A to option B within the confines of a predetermined action arena. Those action arenas are the institutional parameters that define what those options will be. Transformation implies a metamorphosis. It implies that we need to have different options by creating entirely new social structures—new action arenas—that will define entirely new sets of options. By creating new institutions the broader system within which they are subsumed will evolve. The central problem with the transition movement is not that it lacks rigor or accuracy or credibility, it is that it all too often overlooks a central fact: the devil is in the details of the rules and norms that we live by.

In this sense perhaps we could all do a bit of introspection. By introspection I mean a kind of subjective, soul-searching to really observe our thoughts and desires, where they come from, and what they mean. In an economic sense, this means taking a long internal examination of the origin of our fascination with money and what we hold as a true notion of prosperity or justice. For nearly everyone in our society, the word "prosperity" suggests high material standards of living. If we want to define it in any other way, it will take a lot of cultural boxing matches and bruises to get there. The implication is that institutional change is going to be an uphill battle, but the climb will not be nearly as steep as the one we are currently on which is fraught with contradictions and despair and potentially tremendous violence.

CHAPTER TWO

ECONOMIES IN TRANSITION— A SYSTEMS VIEW

As we read the stories of the transition movement, we learn about a great many fabulous projects on community resilience. Local energy descent action plans, workshops on cob ovens and gardening, green building, and other skills how to live more lightly on the planet are crucial. To be sure, these projects will be part of our long-term adaptation to a world without oil. But the crises of resource depletion and global warming that are being confronted by activists in the transition movement are monstrous and systemic. Habitual and continuous economic growth is hastening the decline of our planet's resource base and ecological integrity, but it is also firmly hinged to the underlying structure of our economic system. That is, it is a systemic problem. So is the increasing volatility of our financial systems that is creating economic turbulence worldwide. These crises continue to unfold and are growing in magnitude. If we are going to survive them, we will have to work toward a much deeper transformation of our core economic institutions with the vision that these transformations will bring positive changes to our economic system.

What is possibly the most sobering reality about such systemic crises is that systems cannot be fixed. As much as we would like for it to be so, economic systems are too complex to be repaired or overhauled like lawnmower engines. The concept of *fixing* just doesn't apply. We cannot fix Wall Street instability with legislation, and we can no more fix the relentless drive for economic growth and the consumer trends that go with it any more than we can fix human culture. These are systemic problems can only be solved by evolving and the same is true of the collective human consciousness.

With conscious reflection and appropriate action, each of us can play a positive role in influencing the direction of this evolutionary process. To do so, however, we have to first accept the inevitability of change. Significant change. Every historical epoch has had its beginning and end, and ours is no different. The Oil Age will come to a close and as it does, so will our habitual ways of life to which we have become so attached. The implications of this are perhaps more serious than most of realize or want to admit. The end of the Oil Age is not going to be limited to merely shifting to a new energy source to power our cars. It is going to change everything: industry, culture, human relations, and politics. It is hard for us to see this or recognize this right now because we are so entrenched in our energy-intensive way of life, but this way of life cannot be sustained.

Not many of us want to think about this. In the same way that we don't like thinking about aging and death, we tend to resist the idea that the circumstances under which we live must change. But this resistance creates an unnecessary tension between how things actually are and how we would like them to be for all time, and that tension can make us neurotic and even delusional. Just as we have to accept that our own bodies will change with age and that our lifetimes are finite, so we must accept that our economic system is going to change. Rather than neurotically resisting change, perhaps we could embrace it and become the agents of it. If we accept and embrace transformation, we can gain some control over its direction and development.

Resistance aside, the reality is that the economic systems of the world are already being forced to change. Global warming and resource depletion have been proceeding gradually and have finally reached the point where they shifted from being potential problems to real time problems. Principal among them are the limits to economic growth they are imposing, particularly the global condition of a declining resource base. As this is now being felt, it is creating immediate and acute financial instability everywhere. Our financial systems simply are not geared to function in a world without expanding markets, revenues, and profits.

Financial instability goes beyond limits to market expansion and profits. The boom and bust market bubbles in housing and mortgage derivatives that brought the financial systems around the world to their knees in 2008 and 2009 were not just random events like unexpected snow storms. These were and are institutional problems anchored to the structure of our economic system. The banking crisis that began in 2008 was substantively no different from the Enron/dot-com stock market crash that happened some years earlier, or from the East Asian financial collapse before that, or the Mexican Peso crisis before that, or from the stock market crashes of 1987 and 1929. All of these instances of instability stem from the same core fact that financial institutions everywhere have become gambling casinos that foster wild speculation. Financial market greed and speculation have become accepted practices in our financial institutions and they reinforce a broader cultural norm that idolizes financial wealth and the allure of easy money. This too, like the drive for continuous growth and consumer spending, is a systemic and cultural problem and cannot be fixed with government legislation.

Part of the reason for this is that the governments that would be passing reform legislation are partially responsible for causing the instability in the first place. Our federal government has become so deeply entangled in a web of campaign finance, power brokering, and cronyism that our society has come to be controlled by an oligarchy of corporations and their client agencies within the federal government. Again, this is a systemic problem. Expecting the government to fix this kind of problem is like expecting a dragon to slay itself.

The evidence that this system is crumbling on itself is getting clearer and huge changes are coming our way. This still may be hard for most of us to see. As we look at our glittering, high-tech metropolises with cars and trucks zooming about in every direction, it is difficult to imagine that this must give way to something quite different. Yet the problems of chronic high levels of unemployment, tremendous instability in our financial system, and the growing mountains of unsustainable debt the instability has created are undeniable. These crises are unfolding on multiple fronts and political and established economic institutions are unable or unwilling to deal with them.

Facing these problems can obviously seem overwhelming, but we cannot allow ourselves to become paralyzed by that. Hopelessness and despair are human emotions that stem from a kind of blindness to the opportunities for real and positive change that are now presenting themselves. Rather than surrendering to powerlessness, we should become engaged in our own communities and welcome new approaches to meeting our needs. If we embark on the project of institutional renewal as an extension of the transition movement, this will be a great source of empowerment. As we change our institutions we change ourselves, and at the same time, as we change ourselves we are transforming our institutions. As we collectively change ourselves and our institutions in positive ways, we will influence the direction our larger systems will take. We are the true agents of positive change.

We Are the Agents of Change

The early 20th century American philosopher John Dewey argued that participatory democracy is more than just an exercise in political governance. He saw that an individual's personal development is intimately connected to the creation of a democratic society. For Dewey, people are naturally endowed with an instinct to actively shape the world around them. As we follow this instinct and become engaged in the democratic process, we can create better institutions that advance our social intelligence and foster the development of education, art, health, technology, and purposeful economic activity. By doing so, we create the opportunities for individuals to develop their own unique powers and capabilities. These powers and capabilities, in turn, will allow a person to become an effective participant in the life of the community, and can advance the project of building new and better institutions. In this way, communities and the individuals who live in them can coevolve. In this process we can break the chains that have kept us dependent on this collapsing system and become more self-reliant. This empowerment that comes from being a willful agent of change is the essence of true participatory democracy.

But there is a shadow side to Dewey's philosophy. If we become cynical and turn our backs on democratic participation, the project of building better social institutions grinds down, the fabric of community becomes frayed, and we lose our sense of the common good. Things fall to pieces.

Dewey's message will be increasingly important in the years to come. Like ancient Rome in its twilight, our current system that is built on an unstable foundation of fossil fuels and the idolatry of money is in decline. As it passes its time of glory and gives way to something else, we will have an opportunity to begin making our system into something entirely different. That is, the direction and form of our great transition will be up to us to determine. Our current lives are the link between our past and our future. For those involved in social movements or working for social change, the past is what we are changing from and the future is what we envision changing into. This implies that our work for social change requires that we both look inward into our habits of thought as well as outward to our community. Cultivating a vision for change means questioning everything, including our values, mindset, attitudes and our expectations about what our economic institutions are supposed to do. If we cling to our old habits of thought and action, then we'll continue to drift toward more destruction and instability.

Being agents of change requires that we not only question and break out of our habitual ways of acting and thinking, but also that we be open to what now seems unfathomable. We evolve by stretching ourselves outward and questioning all that is familiar and comfortable, and by working through experiences that challenge our intellect and abilities. We are in a continuous state of becoming and that implies an ever-expanding consciousness. As Dewey saw it, being engaged socially is the way we develop personally.

The real promise of political engagement will not come from another stimulus plan, bailout, or more "sweeping" pieces of Congressional legislation. At least not for some time. Nor is it limited to putting pressure on local officials or businesses. The real source of progress will come from creating something new. It will come from a network of grassroots activists, citizens, and visionaries who are willing to start building new, decentralized institutions at the community level where people have the best chance of democratic control. Buddhist sociologist Ken Jones described such a vision as:

> A working ideal for society and its organizations, in which we are brothers and sisters in mutuality. The network of autonomous groups is now widely regarded as a more appropriate response to many task situations than the traditional model of hierarchal bureaucracy. Economist E. F. Schumacher proclaimed that "small is beautiful," yet the problem remains of effectively managing and coordinating extensive networks in the larger interest with the coercion of a "free" market or a centralized state. The answer for such a commonwealth must surely lie in a high level of public-spiritedness[11]

Perhaps the most important outcome would be that as people become active in their communities and develop institutions that will allow them to become more self-reliant, they can break free of the dysfunctional and dependent relationship they have with their corporate overlords and their vassals in Washington. Once we have demonstrated that we no longer need them for jobs or stability in our livelihoods citizens can build a power base in their communities from which they can demand more accountability from their political leaders. Maybe then politicians and lobbyists will lose some of their hubris and become more accountable.

Obviously this is not going to be easy. Though once enough people have done the work of reflection and soul searching, have developed the skills to research, communicate, organize and so on, the process of building new institutions will take on its own momentum. David Ehrenfeld is quoted in Rob Hopkins, *The Transition Handbook* (2008) as saying, ". . . our first task is to create a shadow, economic, social and even technological structure that will be ready to take over as the existing system fails."[12] Or perhaps another way to say this is that as the current system declines, sprouts of new life will be ready to grow as they are emerging everywhere. If this project of building a new shadow system, brick by brick, institution by institution, gains enough momentum, it will constitute a constructive form of agitation and this will hasten the evolutionary process.

As I said earlier, complex systems cannot be fixed. They do, however, evolve through a series of disruptions and adaptations. The evolution of economic systems is a gradual but uneven process. There are often long periods of relative stasis in which there are fairly gradual changes, but pressures are building in the system. Then there is a sudden flurry of changes caused by a torrent of disruptions and agitation and the evolutionary process leaps ahead. This uneven evolutionary process is what paleontologist Stephen Jay Gould referred to as "punctuated equilibrium."

Perhaps we can be part of the torrent of change. And perhaps we can see our work as the healthy, positive forces of agitation before energy descent punishes us with extreme punctuations. No matter how healthy or positive this agitation may be, it will clash with the established order of things. Of course that also means we agitators could be scorned and ridiculed as rogues or renegades. Then perhaps renegades are what we must willfully become. Hasn't it always been the white crows and black swans who lead the evolutionary shifts? Galileo, Van Gogh, Darwin, Marx, Anthony, King. Scorned in their lifetimes, but remembered afterward. However we define it, as we do this work of creating new alternative institutions we are sending perturbations into what is already a failing system. With time, these perturbations will give direction to the system's overall evolutionary change. The agitators from our past, with their creative minds and tenacity, give us hope that we can evolve.

It is vital to stay hopeful and not to get overly frustrated or impatient in this process. We need to be reminded that the human species is continually evolving—from being to becoming. Much of that process is fraught with difficulty and conflict. When we reflect on the rights and protections that people fought hard for such as the abolition of slavery, banning child labor, securing the rights to vote or collectively bargain, or protection for our water and forests, it is important to note that others were fighting against these changes. The political, cultural, and systemic forces that resist change are always more powerful as they have the advantage of being the established way on their side—the status quo. But if activists had always become discouraged by this then most of us ordinary citizens would still be sweating in servitude, poverty, and stripped of any democratic enfranchisement. We will not know for certain whether we are creating something better until we get there. For this reason, it is doesn't help to be overly preoccupied with the end results of our current action. We must live and work, mindfully, in the present moment and trust that we *are* agents for positive change.

But as we mindfully create new institutions we need a sense of direction. We need to have both a clear understanding of what exactly institutions and economic systems are, and, most importantly, an unflinching view of the nature of the dysfunctional and suicidal system that we are transitioning from. We need to understand the nature of the beast and this requires a systems or holistic view—a sense of the big picture.

Taking the Bird's Eye View

To bring about real agitation-inspired change in our economic system, we need to know what we are up against. It is not enough just to focus on specific actions taken by a particular business or choices made by individuals in isolation. Though these actions and choices are important, a genuine understanding of economic transformation requires a broader, holistic view of the interconnectedness of economic institutions, people, and their habitats.

The general premise in holistic thinking is the notion that the whole is greater than the sum of the parts. In all physical, social, and ecological phenomena there are whole entities or structures that, when they emerge, transcend the basic elements from which they are made. That is, they have emergent properties with their own unique characteristics that cannot be determined from the building blocks that comprise them. For example, if we examine a hydrogen atom in isolation, though the atom is itself a whole, there is nothing about it that can lead us to understand the essential properties of water. It is only when the hydrogen atom forms an interconnection with another hydrogen atom and an oxygen atom that the properties of water will emerge. The properties of water, its buoyancy, freezing temperature, etc., could not be defined by a single hydrogen atom alone.

Economic systems, or social systems of production (SSPs), are the same way. They have emergent properties that cannot be determined from the behaviors of individual people or businesses. As people interact with each other economically through production and exchange and through their interaction, their behavior falls into discernable patterns. These patterns of behavior become habitual and form social institutions. These institutions are formations that eventually take root and evolve along with others to form networks of institutions that cohere into higher-level social systems of production, and SSPs are embedded within broader cultural formations, and cultures are embedded within an ecological habitat. So the complexity of our world goes, like an endless series of Russian dolls (see the image below) and each level has its own unique emergent properties. Such holistic awareness is fundamental to an ecological consciousness. Though mainstream economics largely ignores this as it is still trapped in 18th and 19th century superstitions, it is nonetheless fundamental to understanding economic systems.

Holistic Systems

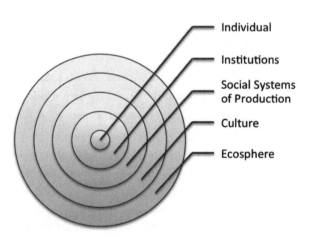

Individual

Institutions

Social Systems of Production

Culture

Ecosphere

It should also be emphasized here that whole formations can be healthy or they can be pathological and destructive. A healthy cell is a healthy formation, but a cancerous one is pathological. A vibrant and productive community that works according to sound principles of justice, ecological permanence, and stability is a healthy formation. An economic system that is fraught with greed, self-aggrandizing consumer behavior, and rampant growth that destroys its habitat is like a cancerous system and if left unchanged will destroy everything around it—including itself. The problems of resource depletion, climate change, and economic instability are pathological systems conditions in the sense that they cannot be understood by merely looking at individual behavior. We can only genuinely understand them when we step back and see how our institutions mold individual behavior in ways that conform to the needs of the overall system.

If we take the long view of the US economy which cranks out about $14 trillion worth of goods and services each year, at some point we become compelled to ask the question: Does this system in its totality have a purpose? We could also ask the same question of the European Union, China, and every other major economic system on the planet. The answer for all these systems is the same: their purpose is to grow. Over the last two centuries the US economy evolved into a kind of super-nexus of corporate, financial, government, labor, market, media, and other institutions that coexist or cohabitate. Those institutions coevolved into a kind of mutual support network or team that collaboratively serves the purpose of ongoing economic growth and expansion. This exerts enormous influence on our surrounding culture and is the principal cause of resource depletion and global warming.

This is a systems condition and systems conditions cannot be fixed. We can, however, work to alter its course of evolution away from the pathology of growth for growth's sake toward something different. Recall that one of the principles of permaculture is to design systems that value edges and margins. The edges or points of contact between the components of a system are important. This is where the action of interaction among the components of a system takes place. In economic systems the edge or points of interaction between an individual person and the overall system are the specific economic institutions. Corporate, media, government, and Wall Street institutions are the medium through which so much of our behavior is structured so as to conform to the overall systems' imperative to grow. If we start building up new and different local economic institutions in small steps, we will not only be creating new rules, norms, and shared strategies to live by in our communities, we will also be sending perturbations into the current pathological system. With time these perturbations will move everything onto a different evolutionary path. In this way, local institutional renewal is vital for both local economic and community renewal and for systemic change. This is where our work is cut out for us and so we need to do some preliminary homework on institutions to create a social system of production for the common good—a commonwealth.

Recently I was having a dinner conversation with a colleague who teaches environmental sociology. When I told her that I see institutional transformation as the way to deal with our mounting economic and ecological problems, she recoiled and said, "People are going to resist you on this. It's just too scary." I didn't disagree with her on that point, but rather emphasized that such hesitance is precisely why these problems—after decades of warnings and scientific proof—are not going away and are in fact getting worse.

I am used to seeing people's eyes glaze over when I mention the word "economics." But when I mention the word "institution" the reaction is typically a faraway and disconnected look, like I am referencing some arcane fact or speaking in a foreign language. The general discomfort so many of us have with talking about institutions might suggest that we associate them with mental hospitals or prisons or government bureaucracies. It might also be true that many of us don't like thinking about institutions because we associate them with a loss of freedom. Institutions, after all, are the primary mechanism of social control. But they are also an intimate, vital, and ordinary part of human life.

Institutions are social structures that prescribe human behavior. The prescriptions come in the form of rules, norms, or shared strategies that organize our interactions with one another. When we talk of human behavior being molded by institutions, we are referring to people interacting within rule-structured situations. And these structured situations that control social behavior do in fact deprive people of certain freedoms. Many of us don't like to think of ourselves as deprived of freedom or institutionalized in our behavior, but institutions are an integral part of everyday life even though the words "institutionalized behavior" carry a heavy, negative connotation. Prescribed social behavior does not mean that our behavior is robotic or preprogrammed, rather it means that there are institutional parameters placed around the choices we make.

Institutions create the dos and don'ts of social situations. Within a family institution, parents create rule-structured situations that place parameters around the behaviors of their children—do this, don't do that. Teachers make students do math in educational institutions even though many kids would prefer not to. Some adults would prefer the freedom to drink and drive, to rob the local grocery store, or physically assault their obnoxious neighbors, but institutions of law enforcement and government place rule-structured parameters around them. Even when we do such ordinary things as going out to dinner we are engaged in institutionalized behavior. How we are greeted when we enter the restaurant, ordering from a menu, our table etiquette, how we taste the wine, and customs for tipping are all prescribed, norm-structured behavior. Neighborhood associations place norm-structured parameters around residents, sports leagues place rule-structured parameters around teams and players. Without social institutions we could not raise or educate our children, protect ourselves from violence, enjoy a good meal at a restaurant, or have any of the social things that make life workable or enjoyable. In this sense it is very good for us to be institutionalized, as otherwise life would be chaotic, violent, and inhumane.

Economic institutions are equally important as life would be nearly impossible without them. Without economic institutions we would not be able to organize and coordinate our work, nor would we have a stable monetary base for exchange or markets. Economic institutions make it so that we are not destined to live in a kind of economic pandemonium and chaos. Walton Hamilton, an early 20th-century scholar who coined the term "institutional economics," defines a social/economic institution and its connection to the higher level system in this way:

> Institution is a verbal symbol which for want of a better describes a cluster of social usages. It connotes a way of thought or action of some prevalence and permanence which is imbedded in the habits of a group or the customs of a people . . . it is another word for procedure, convention, or arrangement; it is the singular of which the mores or folkways are the plural. Institutions fix the confines of and impose form upon the activities of human beings. . . . The world of [economic activity], to which imperfectly we accommodate our lives, is a tangled unbroken web of institutions.[13]

This unbroken web of economic institutions is the social system of production. Each economic institution—household, government agency, market system, corporation, financial, and labor—is collectively integrated into the system. It is through this nexus of institutions that virtually all aspects of our economy are determined: industrial relations, collective bargaining processes, corporate governance, accounting standards, market structures, legal and juridical concepts of fairness, government policies, and consumption behavior.

Though economic institutions create rule-structured situations, the rules are not always fair, just, or healthy. Institutions themselves, like any kind of formation, can become pathological. Slavery is a pathological economic institution that violates basic human rights. So are Mafia organizations that foster violence and hatred, or financial institutions that foster speculative greed, or corporate institutions that foster self-aggrandizing consumer behavior and rampant growth that destroys our habitat like a spreading cancer.

Also, within any social system of production there exists a plurality of interests: buyers and sellers have diverging interests in product markets; labor unions and businesses have diverging interests in contract negotiations. But within this plurality certain structures of power can emerge and dominate the entire system. On this, Herbert Marcuse writes:

> Advanced industrial society is indeed a system of countervailing powers. But these forces cancel each other out in a higher unification—in the common interest to defend and extend the established position, to combat the historical alternatives, to contain qualitative change.[14]

Containing qualitative change means vested interests keep the system from evolving toward something more enlightened so as to maintain a status quo. In other words, in a formation in which there is a plurality of institutions, one particular kind of institution can rise to a position of dominance and control in order to make sure that all other competing interests are subordinated. In our current situation, this would be the large Wall Street and Fortune 500 corporate institutions that defy democratic accountability.

Democracy, as Dewey envisioned it, is empowering individuals with control over their destinies. This control is stripped away by centralized structures of power whether they are totalitarian systems of state socialism or multinational corporations, or an oligarchy of a corporate-government complex. In the last few decades the US economic and political landscape has undergone a transformation. Virtually every major industry including banking and finance, telecommunications, manufacturing, pharmaceuticals, retail, etc., has shifted from a relatively competitive market structure to a fiefdom dominated by an elite club of massive corporations. The large corporation is the central bastion of power and is determined to build global empires. To that end, nearly all of our other social and political institutions have been conscripted.

Playwright and former President of the Czech Republic Vaclav Havel summarized the totalitarian nature of centralized structures of corporate power:

> Enormous private multinational corporations are curiously like socialist states, with industrialization, centralization, specialization, monopolization. Finally with automation and computerization, the elements of depersonalization and the loss of meaning in work become more and more profound everywhere. Along with that goes the general manipulation of people's lives by the system (no matter how inconspicuous such manipulation may be), comparable with that of the totalitarian state.[15]

The source of this institutional power of course stems from the fact that these businesses control the majority of economic production, wealth, finance, and flows of income. But their power goes even deeper than that.

Institutional economist William Dugger has shown that corporate hegemony is maintained through four social mechanisms: contamination, emulation, mystification, and subordination. Contamination places corporate values and motives into noncorporate institutions. More and more the lines of distinction between nonprofits, cooperatives, or educational institutions are blurred. Administrators in our schools and universities are less resembling educators and acting more like corporate managers. With corporate emulation, CEOs are portrayed in the media as heroic cultural figures who are idolized by government and other noncorporate leaders. Mystification covers the corporate agenda with propaganda and apocryphal stories of how the magic of deregulated markets leads to social progress, advancing prosperity, and even ecological sustainability. This has become the paramount role of our print and electronic media. But the aspect of Dugger's insights that captures much more attention is the mechanism of subordination such that virtually all the institutions within our social system of production are used as means to corporate ends.

Evidence of this subordination abounds. The corporate-Washington alliance is a formidable structure of power, but within this structure it has become clear that large corporate leviathans are the sovereign overlords and federal government agencies are their vassals. All the branches of the federal government seem to be preoccupied with running errands for big business. In the last few decades particularly, they have passed legislation, given executive orders, and made Supreme Court decisions that have allowed large corporate enterprises to merge, conquer markets, and establish the kind of monopolistic structures that we haven't seen in the US since the 19th century. In other words, the federal government, with little exception, has adopted the corporate agenda as its own official game plan.

The servile role of the federal government came into full view when the economic crisis exploded in the fall of 2008. The infamous too-big-to-fail corporations from the banking, insurance, and auto industries descended on Washington like Godzilla demanding bailout money on the threat of creating a much larger crisis if the money were not forthcoming immediately. Very few in Washington raised questions about the dangers of being blackmailed by these large companies or about the practice of extorting public money on the threat of creating widespread economic instability. Instead White House officials and members of Congress went scurrying in all directions to comply with the demands of these companies and the cost of this compliance has been truly staggering. The revolving door between Washington and corporate boardrooms, industry insiders holding key positions on regulatory commissions, and the presence of legions of lobbyists swarming the US Capitol are now more palpable than at any time in recent history.

Officials from both parties have rewritten laws on corporate mergers and have pushed through legislation to ratify multilateral trade policies that gave big business the opportunity to build empires internationally. They created tax advantages and subsidies for transnational corporations, revamped environmental regulations to make it easier for businesses to trash forests and blow up mountains, appointed industry insiders to serve as key members of federal regulatory commissions, lavished no-bid appropriations contracts on their friends in business, and both parties have shown an amazing eagerness to make enormous funds available for lavish corporate bailouts when things went sour at the bottom line, particularly the largest players on Wall Street. The recent health care reform legislation that was originally intended to make health care more affordable and accessible did neither. It did, however, amount to a potential massive government subsidy for insurance, health care, and pharmaceutical companies.

The corporate sector is the prime mover and central planner of our social system of production. Our habits of thought have been conditioned for us by these powerful institutions and most Americans are not even aware of it as they accept the corporate agenda uncritically even when they think they're not. I will return to this in more detail in a later chapter on what is conventionally considered ecological, green, or sustainable economics. But as we see this from a holistic view, corporations more than any other institution can exert a more powerful influence on thoughts, beliefs, technology, and culture overall. Does this mean that the individual person in this system is nothing more than a brainwashed corporate automaton?

What About the Individual?

Let's return to Dewey's individual-as-activist model of behavior. Dewey was critical of the notion that was popular in his day that individual human behavior is essentially passive and emerges only in response to external stimuli. Rather he saw that the wellspring of human behavior is a much deeper existential drive to be a proactive force in the world. Dewey writes, "In truth man acts . . . he can't help acting. In every fundamental sense it is false that a man requires a motive to make him do something. To a healthy man inaction is the greatest of woes." Although he did see that the specific action people take is a product of social conditioning. Thorstein Veblen, the grandfather of institutional economics and a contemporary of Dewey, concurred that a person's action stems from an innate tendency to be a seeking, creating, and expressing being. Human behavior, as Dewey and Veblen saw it, arises from what seems to be a universal tendency for proactivity. But the specific things people do are contingent on the social and cultural milieu. We are compelled instinctively to speak, but our language and manner of speech is a social construct.

Veblen probed the fault lines between nature and nurture in social behavior. After gathering much historical and anthropological data on human economic behavior, he identified two general behavioral patterns with respect to economic production and consumption: creative and healthful or predatory and destructive. Both seem instinctive, but which trait actually becomes socially manifest is dependent on certain social circumstance and the kinds of institutions that mold people's actions. If members of a particular community are surrounded by institutions that foster healthy, productive behavior, then it is likely that they will produce and consume in wholesome ways. But if the institutions promote greed, predation, and violence, then it would be no surprise that people produce and consume in ways that are acquisitive and predatory. Sadly the latter circumstance is altogether too common. Our history is an epic story of predatory conquests, enslavement, empire building, and ravaging ecosystems for little purpose beyond what Veblen identified as conspicuous consumption.

These psychic and social forces that compel human behavior are of great interest to Buddhist scholars. For Ken Jones, the source of pathological behavior exists in a deeper existential "hunger" or "psychic greed" for meaning and self-identity. This hungering originates in a sense of hopelessness and a primal attempt to fill a kind of emptiness or void. Jones writes:

> Liberation from our sense of lack is impossible as long as we evade accepting it's ultimately ego-created nature and instead go on trying to fill the imagined hole by top-loading our lives and the world with ego-affirming behavior that is needless, ineffective, and destructive.[16]

Taking these ideas together it would make sense to say that deep within us all are the primal seeds or potentialities to be creative and productive along with a primal desire to be fill an insatiable hunger for self-aggrandizement. Which seeds get watered and grow into patterns of behavior depends largely on what is going on around us. One or the other can become habitualized into deeply entrenched social institutions. It would appear that the seeds of destructive behavior have been cultivated and deeply woven into our institutional and cultural fabric. Consumerism and conspicuous consumption are everywhere as are Wall Street speculation and the idolatry of financial wealth. And as ecologically destructive and destabilizing as these types of behavioral patterns are, they have in many ways become cultural norms.

It is not hard to see how the institutions of modern capitalist societies have normalized destructive, ego-aggrandizing behavior. Businesses are constantly watering those seeds of insatiable hunger, predation, and desire with advertising. Financial institutions, once created for the purposes of economic development, are little more than casinos that not only foster greed, but normalize it. Ken Jones contends that ". . . capitalism is a system that rewards greedy and acquisitive behavior, and it legitimizes the harshness and violence associated with gross disparities in wealth and income distribution."[17]

Buddhist philosopher David Loy sees that the US economic system, "institutionalizes greed in at least two ways: corporations are never profitable enough and people never consume enough . . . that media companies never question the delusions spawned by their manipulative advertising, which are specifically designed to foster consumerist impulses to buy things."[18] Echoing these sentiments, Buddhist philosopher and social activist Sulak Sivaraksa argues that ". . . the message from the media is that happiness is something that can be gained from consuming endless quantities of stuff."[19] Similarly, Zen philosopher Thich Nhat Hanh sees corporate media as a source of mental and spiritual pollution that encourages greed, violence, and anxiety. He sees no distinction between the pollution of consciousness and the destruction of our natural environment.

Corporate media is an enormously powerful institution. It instills vain and capricious values in large numbers of people who perhaps otherwise would not have a predilection for self-indulgence. It compels people to consume less intelligently and gravitate toward flashy and ostentatious products rather than things that are wholesome, genuinely beneficial, or reliable. Corporate advertising exploits prejudices and yearnings for no other reason than to sell and expand market share with little or no ethical consideration or concern for public morality. Ads imbue the public with unrealistic ideals for what is considered "the good life" or the "American dream" and condition all who are under their spell to accept that the perfect consumer is a person who is indifferent to the environmental damage consumerism does and has no higher purpose in life than to consume ever-growing quantities of stuff. Large and powerful corporations use media advertising to create epidemics of neurosis in which seething masses of otherwise psychologically healthy individuals are turned into addicts and what Buddhists call "hungry ghosts."

In the Buddhist tradition, the realm of hungry ghosts is a place where everyone has enormous bellies, but very narrow throats. It's a kind of hell in which they are all besieged by an unquenchable thirst, but when they drink the liquid turns to fire making them thirstier. The hungry ghosts are overwhelmed with hunger, but the narrowness of their throats prevents them from swallowing food. So they suffer from ever intensifying cravings. The message of the parable is that conspicuous consumption has never brought true satisfaction and instead creates more dissatisfaction. Nothing could be more helpful for sales and marketing than a critical mass of dissatisfied, insatiable consumers neurotically looking for more.

What makes these problems so intractable is that they have a way of self-reinforcing. The primal impulse for ego-aggrandizement is encouraged through a barrage of media advertising and this leads to patterns of consumer behavior that become habituated as a cultural convention—consumerism becomes institutionalized. In another sector, the primal impulse of greed and acquisitiveness gets encouraged by get-rich-quick schemes on Wall Street, which also lead to predictable patterns of behavior that have become conventional—greed becomes institutionalized. All of these behavior patterns and institutions become diffused into our social system of production and culture, which become mirrors that reflect these habits back to the population. Our culture conditions our way of thinking, which further conditions our behavior, and so on in a cumulative process. It is no wonder that environmental destruction and financial market instability have become so enormous that they appear to have no solution.

Returning to the original question about the individual, is each of us merely a hungry ghost unwittingly trapped in the maws of Fortune 500 corporations and helplessly drifting toward our own destruction? Perhaps many of us are, but it doesn't have to be that way. It is true that the institutional forces around us run in powerful currents, but we are nonetheless active and volitional beings and we have the capability to ignore corporate advertising and behave differently if we choose to. If we create different habits that become embodied in new institutions, those institutions will become the new mirrors to reflect those new habits to us. Echoing Dewey and Veblen, one of the founders of modern systems thinking, Ludwig von Bertalanffy, tell us, "[S]ystems theory sees the individual as primarily active ... Man is not a passive receiver of stimuli coming from an external world, but in a very concrete sense *creates* his universe." Since it is in our nature, we are indeed capable of creating a new universe—thought by thought, habit by habit, institution by institution, and community by community.

Creating a New Universe

Our great transformation will not be an attempt to go out and completely overhaul our economic system or culture. That is not possible. It is rather an evolutionary process of change in which we use the most important tools we have available, our conscious minds and will, to transform our habits of thought and action in the direction of something new. Perhaps the best way to begin our great transformation is by opening some space for thinking creatively about the kinds of institutions we will construct and how those institutions will weave together into our shadow system.

We should also be asking questions. Many questions. What is the central logic followed by established institutions? Do we want to follow that same logic? If so, then how can we claim to be building something new and different? If not, in what ways are our new institutions genuinely different? And how do we make ourselves different? What are the key factors that make those new institutions cohere into a new and different system? What makes that shadow system successful in achieving its purposes? What makes it fail? What would constitute the most important institutions of our new system that would be foundational social building blocks? What are our goals? How would we make our shadow system cohere into a stable structure? What are the potential pitfalls? What is different about our new system that will allow it to succeed where others have failed? What exactly is the measure of that success or failure? How is that measurement consistent with our values? And what are our guiding principles?

As we strive to answer these and many other questions, we'll get a better sense of how to proceed with institution building in our communities. Such work is the essential task of economic re-localization. Each entity that we create or re-create—household, social enterprise, nonprofit, cooperative, government agency, financial institution, market exchange, guild, or labor organization—will itself be a complete and semi-autonomous module existing in a state of plurality and mutuality. These modules, if assembled together mindfully, can become the institutional dream team comprising a new commonwealth.

One problem to be worked out in this process is how to effectively coordinate each of these institutions with a shared set of values and principles so that they can raft together in a mutually supportive network. This requires praxis—practice that reflects those core values and sound principles. If the work is done reflectively with a high degree of public-spiritedness, each institution will naturally cohere into what institutional economist John Commons refers to as new "social collectivities." This network or friendship of institutions will be establishing new social conventions, working rules, and a new local culture which itself will gain a life of its own.

When we refer to culture here we are not limiting our conception to language, music, food, art, etc. Culture in the institutional sense has a broader meaning as a constellation of what Lewis Mumford refers to as both material and nonmaterial "technics." Material technics are all the physical things that characterize the communities such as: buildings, tools, technology, clothing. The nonmaterial technics are the social practices: language, music, customs, habits, conventions, mores, etc. Together these technics form the web of life in the community. But perhaps most important of all is that this work will be empowering in the Deweyan sense of democratic social participation.

In time and with some luck the fruits of these efforts will be to shake off the ties that bind us to the established dying and dysfunctional system—a system that spawns pathology of mind and pathology of action. Our great transformation is the inward work of changing ourselves and at the same time is the outward work of changing our communities. Here are some possible goals for our great transformation:

- To transform households away from the conventional role of acquisitive and obsessive consumers to become integral parts of an ecologically and economically stable community.
- To transform financial institutions away from speculation and the expectation that financial systems are always going to pay off exponential returns; to restore financial institutions to their original and core function of providing financial services for local development.
- To transform business models away from growth obsessed, profit maximizing corporations to social enterprises that exist to provide for the needs of the community.
- To transform markets and market exchanges away from a nexus of greed and speculation to community centers in which people not only make necessary exchanges of money, goods and services; to a community marketplace that is the center of a vibrant community culture.
- To transform local governments away from the conventional agencies that are captured by moneyed interests to agencies that are democratically accountable to their communities.
- To transform labor organizations away from conventional models of exclusive self-interest to the stewards of craft traditions and the joys of good work.

With these transformations, re-localized communities will become the sprouts of new life and healthy agitation that will evolve toward systemic change. People will begin to act and think in different ways; hopefully in ways that will foster stable livelihoods, ecological permanence, and a serious commitment to social justice. In the Buddhist tradition, this would be considered "right livelihood." The "rightness" of it not based on moral judgments or commandments, but rather what is developed with our natural capacity for compassion, wholesome values, and the creative spirit that springs from a clear mind. To do this work we will need what E. F. Schumacher calls a "meta-economic" system; that is, an ethical and ideological basis that will inform the specific aspects of institution building with wisdom and insight.

With clear minds, wisdom, and true values we can begin to envision a process of creating a new universe. To be sure, there is no shortage of cynics out there ready to do some name-calling and make their standard cracks about unrealistic aspirations, etc., but it has always been that way. Just as Diogenes went out in broad daylight with a lantern looking for an honest man, there is going to be someone smirking while questioning our motives. If we were to listen to these people, none of us would ever put oneself out there to write a play, compose music, invent a theory or a machine, start a business, go for a college degree, become an athlete, run for office, or learn to dance ballet. And certainly none of us would work to create a new universe.

This is no time for cynicism, but it's the perfect time to dream up new ways of living. This is no time for predatory moneylending, but it is the perfect time to help organize a new financial cooperative in your community that is created to provide badly needed services—though not for the purpose of making big money, but because it is the right thing to do. This is no time for building another corporate empire that will dominate global markets, but it is the perfect time to create a modern-day agora—a marketplace as community center where we expand the farmers market to a permanently established location with regular fairs, workshops, meetings, poetry slams, live music; and where local guilds practice their trades and sell their wares; an online directory and newsletter to keep the community informed of what is happening and what goods and services are available—handicrafts, food, energy, handyman labor, repairs, finance, medical services, fabric, and so on.

This is also no time for more economic growth for growth's sake, but it is the perfect time to create a commonwealth comprised of social enterprises that provide food, health care, energy, entertainment, and a million other things because the people in the community need them. It is the time to transform pathological consumerism into mindful living, and time to give our ecosystem a chance to breathe again.

Conclusion

Future historians will look back on our current period and chronicle the events: skyrocketing energy and food prices, riots, nuclear power and financial meltdowns, climate change, and total resource plunder. If the period that follows is a Dark Age of the Third Millennium, the historians might wonder why we with all of our intelligence and technology failed to do something about it before it was too late.

For this reason and because of the moral responsibility we have to our future generations we have to be forward-thinking and proactive. The emphasis here is on stepping onto another path before it is too late. The transition movement has given us a head start. But the ideological basis of that movement is such that it shies away from the work of institutional transformation. This will have to change if the transition movement is to actually realize its goals of creating new community-based systems that have the resilience to adapt to a world without oil.

The intention here is to build a framework for how we act in the world and how we think about the world. The movement is directed at thoroughgoing systemic change, but the path toward change begins with local, community-based institutional development where people have the most control and can be most effective. It is local action and engagement that is intertwined with a holistic consciousness and an awareness that our every action is linked to every other. More specifically, it is a process of creating community-based institutions that counter the pathology of our established system, which left to its own logic and momentum will certainly bring about a new Dark Age.

Economic institutions are just as integral to human life as any other type of institution. But they evolve and change with time. One of the founders of institutional economics, Russell Dixon, sees economic institutions as simply behavior patterns that have solidified into conventions, customs, and folkways. Solidified, though, this does not mean unchanging. A change in one aspect of culture gets diffused into all other aspects of culture and the entire pattern evolves. Just as biological forms evolve with time, so do institutional forms. The slave systems of ancient Greece and Rome gave way to the feudal manors and guilds of the Middle Ages. And eventually feudal institutions gave way to the market system, the corporation, and the modern nation state. These were large scale transformations and we are now on the cusp of another great transformation in which the old is giving way to the new.

We are at a critical juncture where it is has become very important to ask ourselves if what we are doing is truly new and evolutionary or if it is merely the same old procedures, conventions, and arrangements with new window dressing. In the next chapter I address the problem of current and fashionable green economics, which is largely same old conventions or at best half-hearted attempts at something new, and I'm not just talking about the all-too familiar process of greenwashing.

CHAPTER THREE

THIS ROAD PAVED WITH GOOD INTENTIONS

In his 1966 essay, "The Economics of the Coming Spaceship Earth," economist Kenneth Boulding declared that the image we have of our relationship with the planet is going to change. His message was primarily directed at his fellow economists who tend to see the national economy as a kind of super machine that churns out ever larger quantities of things for us to buy. In a way, these economists are correct. Aside from recessions that occur now and then, our economic system has been running, nonstop, like a machine programmed for continuous growth. But Boulding saw that such a perpetual growth machine will eventually slam against a wall of limitations imposed on us by the laws of nature and physics. This will eventually jar us into changing how we view our relationship with our environment. Boulding concluded that a rather more sobering and realistic image would be to see ourselves as passengers on a spacecraft, which implies that our survival depends on how we use the finite provisions that are stored on the ship.

Decades have now passed since Boulding wrote his essay, but the economics profession remains captured by the myth of the everlasting growth machine. Though economists acknowledge that resources are scarce and finite, they assume that the machine will always be kept running with the support of new technologies. This is like believing in magic or the transmutation powers of alchemy. Ancient alchemists were obsessed with the "elixir of longevity." In addition to their fascination with trying to turn lead into gold, they sought tirelessly to find a substance that could keep restoring the human body to youth so that they could live forever. In a similar way, economists hold on to the belief that technology is a magic elixir that will allow the growth machine to defy the immutable laws of physics and expand forever. Their belief is that as we bring one important resource to exhaustion, we will always be able to concoct new resources with new technology, and when those resources are exhausted, others will be invented to replace them, and so on into infinite growth. Whenever they are called on to address the physical limits to economic growth, economists generally turn to this fantasy in which limitations don't exist and in which people will always be given more things to consume, generation after generation, for all time. Most economists would admit that things like turning lead to gold or eternal life are scientifically not possible, but for them such science apparently doesn't apply to economic growth.

Economists are not entirely responsible for this fantasy. Featured on the cover of *Business Week*, February 2010, President Barack Obama declared stridently that, "We are pro growth. We are fierce advocates for a thriving dynamic free market." Obama was saying this at the time when the Federal Reserve was pumping hundreds of billions of newly fabricated dollars into the banking system trying to speed up the economic machine. Obama reiterated, "The Fed's mandate, my mandate, is to grow our economy. And that's not just good for the United States, that's good for the world as a whole." As we remain relentlessly committed to growth, like the passengers on a spaceship who overconsume their provisions, the end result is certain to be perilous. Yet the leaders of our most powerful institutions are ignoring this certainty. To be fair, though, it may be impossible for them to do otherwise because economic growth is what citizens demand. The tempting idea that material prosperity will always expand is accepted as a truism and is deeply woven into our cultural fabric.

At some point in time we will be forced to awaken from this fantasy. The immutable physical limits to growth will make certain that our practice of spinning endlessly on the production-consumption treadmill will eventually slow to a point at which our most central economic institutions will begin to destabilize—perpetual growth will give way to perpetual crisis. For those of us who are seeking a more stable future the time for awakening is now as we are beginning our energy descent at the end of the Oil Age. As conditions of growth give way to conditions of crisis, communities will be forced to adapt. And as engaged citizens we have the power to develop ecologically sound and economically stable ways of living in the world. But as I have argued in the previous chapters, we will first have to do the hard work of transforming our institutions, our mindset, and our expectations about what our economies are supposed to do. This work is predicated on developing a genuine understanding of the problems surrounding economic growth, not on fantasies and mythology. With this true understanding citizens can gain a more realistic view of what is possible and what is not, and with this knowledge envision a more realistic future for their commonwealth.

The Economics of Growth

It is fair to say that part of what drives economic growth is an expanding human population. The Labor Department estimates that the US economy has to create over 90,000 additional jobs every month just to keep up with the new entrants into the labor force. To create those new jobs, the economy must grow. But the relationship between population and economic growth is a chicken and egg one. As we can say population drives growth, we just as easily say that our increased population is only possible because the economies of the world have been expanding and therefore able to support our larger numbers. Demographic data shows that during economic downturns, population growth rates subside and then pick up again during boom years. The Great Recession showed the slowest population growth rates since World War II and births fell by 7.3 percent from 2007 to 2010. During the Depression years of the 1930s birth rates fell by a third.

Looking only at population numbers is a very limited way of understanding the problem of growth. The environmental stresses caused by increased population are not evenly distributed. As everyone on the planet is chasing after what now seems to be a globally homogenized version of the "American Dream," those who achieve that material standard of living will place much greater stress on the planet than the majority who don't. So the stress on the planet is not measured only by population numbers, those numbers must be multiplied by material production and consumption. Americans have the highest per capita consumption rates in the world, yet seem to have very little concern over their numbers, and are more likely to claim that overpopulating the earth is something that other people do. This population question aside, there nonetheless remains another institutional driving force behind continuous economic growth. Our financial institutions constitute the primary drive wheel behind this growth machine.

Financialization is a buzz term that has been tossed around considerably since the banking crash started in 2008. It basically means that the logic that drives Wall Street is what drives the rest of the economic system. While this is true, it is only part of the story of economic growth. Economies are pushed to grow because that is also what people expect their financial investments to do. Pension funds or any other portfolio filled with stocks and bonds are considered to be performing well when they generate returns that accumulate over time. It is perfectly natural for us to expect continuously compounding financial returns without giving it a second thought. Even if the markets crash, which they do on occasion, the general assumption is that they will bounce back and continue appreciating. "Markets always come back," they say on Wall Street, "always." If we look at the last 200 years of stock market history, that certainly seems to be the case—a long term trend showing a 7 percent rate of return after factoring out price inflation. It is no surprise, then, that people see this as a normal aspect of our economy.

Cultural norms, however, are not always healthy nor are they always grounded in reality. Is it truly valid to expect that just because the market values of Wall Street securities have always grown in the past, they will continue to do so in the future? Mathematician Albert Bartlett notes that growth is an adolescent phase of life and stops at maturity. Is it valid for me to say that since I grew from zero to six feet tall in the first twenty years of my life, I can expect to grow to twelve feet tall in another twenty years? And grow to twenty-four feet after that? Continuing growth after maturity is either obesity or tumorous and neither is healthy. Human bodies have natural limits to growth and so do economies, though in our culture we tend to believe otherwise.

We believe in this because the endless accumulation of money is a very alluring idea. People pour trillions into pension funds, hedge funds, mutual funds, etc., because of the promise that these investments are going to continuously appreciate in value and will provide for us in our retirement, or pay for our children's education. If the grandparents of a newborn child created a $12,000 trust fund when the child was born on the condition that the money was to remain invested in the stock market until the child grew up and retired at the age of 65, all the while earning that annual 7 percent real return, the trust fund would be worth nearly a million dollars at retirement. The child could look forward to becoming a successful millionaire at retirement without ever lifting a finger. In our culture, we say that money doesn't grow on trees; rather it grows in the bank toward an infinite horizon of financial wealth.

By the same logic and cultural norms, businesses expect their earnings to grow and working people expect their paychecks to grow. For all this growth in money and financial wealth to be appreciable or meaningful to people, the amount of stuff you can buy with the money must also grow—real economic growth. After all, what good is it to become a millionaire if you can't buy a million dollars' worth of stuff with the money?

This expectation is a major force for economic expansion. If the money side of the economy is expected to keep ballooning out, it follows that businesses are pushed to generate new sales and create new markets. If they succeed, then the businesses are rewarded with higher stock prices and management with hefty bonuses. The profits businesses make from their new sales provide financing for new capital, which will drive production and sales even higher. For that, though, they need to find more consumers with voracious appetites. In time, this forms into a perfect circle or positive feedback loop. We expect and demand that our financial wealth continue to accumulate, this financial growth is derived from expanding business profits, business profits are derived from expanding sales, and expanding sales are predicated on the creation of a consumer culture that idolizes the accumulation of financial wealth. In this way, economic growth is driven by its own self-reinforcing momentum that can only be stopped by either willfully breaking out of this circle or by hitting the wall of physical limitations—negative feedback. As we remain in denial about all this and willfully refuse to change our habits of thought and action, then smashing into the wall will be a certain outcome.

Hitting that wall is likely to come sooner than we like to think. To gain a perspective on this, consider the basic mathematics of growth. According to the number crunchers at the Social Security Administration, the US economy is forecast to grow by about 2.2 percent annually, adjusted for inflation, over the next several decades. Though 2.2 percent seems modest, and perhaps even conservative, it is an exponential growth rate. Exponential growth means that the things we produce will accumulate like a rolling snowball and at a 2.2 percent rate that snowball will double in size approximately every thirty-three years. So that newborn child with the trust fund should see the US economy grow from today's $14 trillion to $28 trillion in the first thirty-three years of its life, and then to $56 trillion by retirement, an amount that is larger than the economy of the entire world today. And then it will double again to $112 trillion in the next thirty-three years.

Of course the limited and finite resources of our planet, particularly oil, will render this impossible. While every economy in the world is plunging deeply into debt in a desperate scramble to restore economic expansion, the resource base of the planet is moving in the other direction. Long before the

US economy doubles or triples in size, the resources that have up to now been supporting growth will collapse under the sheer weight of it. We will reach a turning point at which the same mathematics of growth will turn into the mathematics of decay. In fact, with peak oil we are coming to that turning point right now.

If we are to learn anything from the various crises we've been facing in recent years—skyrocketing energy prices, oil spills spreading out in the water like giant sea monsters, looming threats of nuclear power plant meltdown—it is that our growth-driven systems are finally reaching their inevitable limits. There are only so many resources we can bring to bear in the production process, only so many toxins we can dump into the atmosphere, only so many markets in which to sell, and only so many people with the purchasing power to buy. These limitations will eventually force us to disentangle ourselves from the culture of growth and the consumerist ideal of the American Dream. And though many of us are coming to realize this sobering reality, our economic and financial institutions, as well as our culture, are lagging behind. Far behind.

If we look at growth data from the last century, it is easy understand why Wall Street would claim that markets always come back to a growth trend. But that century-long trend of growth also coincided with the age of oil—a one-time, cheap, ready-to-go energy source that is evaporating at an accelerating rate. The stark reality is that as global oil production peaks there is nothing in the universe that can replace it, not even with alchemy. And as we expect to sustain infinite growth on a finite planet, our expectation is as foolish as the alchemists' attempts to achieve eternal life.

This is not only foolish, it's deadly. With a declining global resource base, any measure of real economic growth in the future will be zero-sum growth. This means that expansion in one place can only come about by cannibalizing resources from another. The passengers on Boulding's spaceship who recklessly exhausted their resources are likely to begin stealing the last bits of food from each other before facing their certain death. As we remain fixated on economic growth, we will be sentencing our future generations to a perpetual state of conflict and warfare over dwindling supplies of oil, fresh water, and arable land. Such conflict, of course, will only hasten the deteriorating conditions on this planet.

Ultimately, however, Boulding's spaceship metaphor is a bit misleading. Our planet is not just a vessel or warehouse of useable resources, though in our economic behavior we treat it as such. The carrying capacity of the planet is a vast web of life—the inter-being of air, vegetation, water, trees, topsoil, animals, and humans. As relentless growth destroys the fabric of this web, it not only destroys the fragile qualities that have supported us for several millennia, it destroys the very wellsprings of our spiritual nature.

I often hear people say that economic growth will be less damaging because production will be more electronic or digital rather than based on traditional raw materials. There may be some truth to this claim that the future is the digital age rather than the Oil Age, but there are some problems with this line of thought. Sending electronic files back and forth in cyberspace is certainly an identifiable aspect of any modern economy. That activity, however, is a means to an end, and the end is to make money so as to buy material things. The electronic symbols will continue to grow in importance as they have up to now, but most of these symbols are created to augment the material side of our economic activity. Producers of software, designs, or texts are primarily in the business to make money that will be used to buy homes, cars, hot tubs and so forth. It is surreal or dream-like to imagine that all of our oil- and resource-based production will eventually be supplanted by electronic files—a kind of science fiction fantasy world in which people have drifted away from the crust of the earth and into cyberspace. Moreover, according to MIT graduate and inventor Saul Griffith, "The Internet's energy and carbon footprints now probably exceed those of air travel . . . perhaps by as much as a factor of two."[20] The reason is that the enormous amount of activity that we do online is powered mostly by coal burning utility plants.

As we download and store PDF files, send jpegs back and forth, watch streaming videos, trade stocks, pay bills, update Facebook pages, and so on, most of us do not have a clear sense that this digital activity has a physical component and burns prodigious amounts of energy. People everywhere have formed an expectation that they should have access to vast amounts of information and images at their fingertips at all times. We seem to be unaware of just how energy intensive this digital activity is because the primary use of energy in the process is not in our smart phones or laptops, but in the data centers that support them. According to a year-long investigation by the *New York Times*, there are about 3 million data centers worldwide that burn through about 30 billion watts of electricity every year. These data centers are essentially warehouses packed with high-powered servers that process trillions of gigabytes of data every year, and the amount is growing exponentially. A data center designer interviewed by the *Times* noted that one data center uses up about the same amount of electricity as a medium sized town. Part of the energy used is directly to process data, but much more of it is used to keep the servers running around the clock—whether they are being utilized or not—just to keep consumers satisfied with their constant access to everything online. To keep all this going, data centers run auxiliary diesel generators and use lead batteries so that that the servers will keep humming in the event of a power failure. These super computers also get hot, so the data centers also constantly run industrial cooling systems. As noted in the *Times*, "[consumers] drive the need for such a formidable infrastructure." Like every other sector of the global economy, this stupendous growth in online activity will eventually hit the energy brick wall. When that happens, the emerging fantasy that the digital age can free us from physical limitations will be shattered, and the economic and social tailspin that will follow will be incomprehensible.[21]

The danger in this fantasy is that it distracts people from the true material nature of our existence. We are flesh and blood human beings with a deep physical and spiritual connection to our planet. Rather than floating into cyberspace—as people seem to be attempting to do as the cell phone addiction has spread into a global pandemic—we need to be forming a deeper connection to our physical habitat if we want to avoid destroying it. Out of sheer necessity of survival, our focus will be more and more on the material and ecological conditions that surround us rather than electronic symbols. If our world is always framed for us by Facebook, Twitter, and streaming video, we take ourselves even further away from our fundamental material existence—an existence that is completely and utterly dependent on fresh air and water, healthy topsoil, vegetation, and minerals.

Put another way, we cannot eat, drink, or clothe ourselves with video images or electronic symbols, though there is a growing mass delusion that seems to suggest that is somehow possible. It is true that the US economy is much more service oriented than manufacturing, but this doesn't mean we don't consume less manufactured goods or food. It means that production of those things that rely on the carrying capacity of the planet have been shifted offshore and then imported to be consumed.

If we think that we have become less of a resource consuming society the data from material flows analysis tells us otherwise. Even when we are producing more and more electronic symbols as a percentage of our GDP, the amount of material extraction and consumption has nonetheless been increasing to unprecedented levels. Part of the reason for this is what Juliet

Schor in her book, *Plenitude*, 2010, describes as accelerating fashion cycles.[22] The pace with which new products hit the market to replace their older counterparts is speeding up as companies scramble to conquer new markets to stay in the habitual game of growth.

Breaking out of the growth habit will prove to be enormously difficult. The expectation of the eternal accumulation of wealth has become so familiar that most of us have never tried imagining it being any other way, nor would we want to. It is very difficult to convince people of something they don't want to hear. People will fiercely resist any notion that suggests the elevator of riches has finally started its descent. As difficult as this may be, breaking the growth habit will prove to be a far less daunting task than facing the trials that will come from transforming our planet into a toxic waste dump, from condemning much of the world's population to destitution, or from falling into a dystopian society fraught with endless and violent struggles over shrinking resources. The longer we wait to begin our transition to a different mode of economic production, the more the mathematics of decay will accelerate, and therefore the harder it will be for our children and future generations to survive.

Some in the economics profession are beginning to acknowledge these problems, yet they still resist mentioning limits to growth. Limits to growth have been a topic of interest in economics for over a hundred years since Thomas Malthus made his dismal forecast that population growth rates would eventually exceed the food supply. More sophisticated analysis of the relationships between population and industrial growth were conducted in the early 1970s by a think tank of environmental scientists known as The Club of Rome. Their report titled *Limits to Growth* published in 1972 spurred a debate in which their findings were disputed by economists who insisted that the limits argument was ignoring the all-important assertion of technological salvation. As being saved by technology is a popular—nearly religious—belief, the economists seemed to have gained higher ground and continued to kick the concerns about limits to the curb.

Economists are like the last standing alchemists clinging to their dreams of finding the elixir of eternal growth. Their world of make-believe has merged with a modern ecological consciousness to spawn "green economics."

Rethinking Green Economics

Many of us have by now heard stories about business "greenwashing." The stories are usually about companies that make bogus claims that their products are made in ecologically sustainable or socially responsible ways. These are cynical attempts to cash in on the premium prices those products can command in the marketplace as organic or fair trade. The concerns I raise here about "going green," however, are different. With all the best of intentions, many ecologically minded economists are suggesting that the crises stemming from limits to growth can be circumvented by making capitalism greener. The core aspiration is "do good and do well." That is, to make ecological sustainability compatible with conventional economics and business models and save our ecosystem from damage without having to do the real work of changing mindset or expectations. Such "have your cake and eat it too" promises are potentially dangerous because they give people a false sense of what is possible, and they distract us from the more important work that will have to be done if we are serious about achieving any semblance of ecological permanence.

At the core of green economics is the promise that businesses can continue performing for the bottom line as always and achieve ecological sustainability at the same time. For some years now green economics has been presented under various labels such as "The New Economy," "The New Green Revolution," "The Green New Deal," "The Green Collar Economy," "Green Capitalism," or "Natural Capitalism." Whatever we call it, the allure stems from the promise that that going green can be fun, easy, and most important of all, profitable. This is a classic and irresistible win-win proposition that is music to the ears for Wall Street and our consumer society. It is comforting for people to hear that we don't really need to change our way of life or expectations about what our economy is supposed to do. But we are fooling ourselves if we cling to the belief that the very same growth-driven businesses models and the consumer culture that have brought us to the brink of ecological ruin is now the go-to solution for ecological sustainability.

The main proposition of green economics is that conventional business models and markets are resilient enough to solve even the most complex ecological problems. One aspect of this is the formation of capital for private-sector investments in major projects: dazzling technologies such as solar arrays, Stirling machines, algal ponds, wind farms, building retrofits, electric cars, carbon scrubbing, or eco-housing just to name a few. The assumption is that private businesses that make these investments will earn handsome profits and returns, and along with this goes the promise to create many good-paying green collar jobs. The money earned by the greened workers can be spent on the growing market for greener consumer goods, and the businesses that successfully market those goods will be plowing their profits back into greener development. Everybody wins: producers make profits, workers get jobs, and consumers save the planet by shopping. Who can resist drinking this green Kool-Aid?

Perhaps the most articulate recent expression of green economics comes from investment fund manager and green economics advocate, Woody Tasch. In his book, *Slow Money: Investing as if Food, Farms, and Fertility Mattered,* 2008, he calls this approach "restorative economics" using business models that integrate fiduciary responsibility—shareholder returns—with conventional principles of asset management. He argues that asset management can be redefined as care for the integrity of natural ecosystems, as well as to protect cultural and biological diversity. Tasch refers to this as a kind of business realignment of profits with the carrying capacity of the planet. He sees this as an economic and cultural transformation that will reduce the environmental stress we place on the planet, but will find ways to continue with conventional practices including our way of life and a commitment to economic growth, all the while reducing our ecological impact.

With green economics ecological thinking can be reintegrated into what Tasch refers to as "modern notions of entrepreneurship and fiduciary responsibility." In other words, green capitalism. He implores us to revisit the meta-economics of Schumacher and others who view economic activity from a higher purview of moral and aesthetic imperatives. But Tasch retreats from suggesting that we genuinely adhere to those imperatives by claiming that to do so would be ". . . too at odds with modern notions of entrepreneurship and fiduciary responsibility." His aspiration, it would appear, is to find some middle ground between a genuine commitment to reaching our goal of ecological permanence and the practical concerns of generating profits for investors. Though such pragmatism is understandable, it is disturbing to see his deceptive and clever use of language to get his point across.

Throughout his book he refers to "modern fiduciary responsibility" which is really another way of saying "corporate capitalism." But by carefully using the word "modern" and juxtaposing it to the classics of Schumacher and others, the suggestion is that Schumacherian economics and the views of the meta-economists are antiquated. I raise this concern because this is the same kind of rhetorical manipulation Bill Clinton, Alan Greenspan and Robert Rubin used to deregulate global financial markets in the 1990s. Their repeated use of the word "modern" as if it were synonymous with "deregulation" served to frame those who were skeptical of deregulation as curmudgeons. The so-called curmudgeons were thus shouted down, deregulation became the rule of the day and the results that eventually followed were sublimely catastrophic. It is almost like saying that if we call slavery something else, say "modern human labor management," we can go on with a pretense that it is somehow no longer a fundamental violation of human rights.

By whatever name you choose—fiduciary responsibility, fiduciary capitalism, corporate capitalism, or just plain capitalism, there is actually nothing modern about it. Capitalism in its various incarnations has at least a 400 year history. Throughout these four centuries, capitalism as a system has shown very little if any indication that it is compatible with ecological permanence, but the evidence of the opposite is overwhelming. And throughout those centuries, the core principle for organizing economic institutions has been to generate investor returns through market expansion and trashing the planet along the way. Slowing down the pace is not a solution because it isn't consistent with the logic that governs the dominant institutions of capitalism.

Jeanne and Dick Roy, founders of the Northwest Earth Institute, warned about this. The Roys assert that there is asymmetry between the goals of fiduciary responsibility and true sustainability. The result, the Roys write in their newsletter, is "sustainability lite" or a half-hearted approach to sustainability. The Roys quote a typical business-oriented approach to sustainable practice: "I am an advocate of sustainable practices so long as they increase the bottom line. Otherwise, how would I sell this approach to management."[23] Green economics guru Paul Hawken also warned that the allure of having cake and eating it economics is leading to ". . . the dumbing down of criteria and the blurring of distinctions between what is, or is not, a socially responsible company." Nonetheless, Tasch embraces this half-hearted approach. He writes,

> We must be critical but not too critical of the environmental and governance-related failings of mature corporations. We must be critical but not too critical of the environmental and governance-related failings of start-up companies that explicitly embrace the triple bottom line.[24]

Tasch and other advocates of triple bottom line (profits, sustainability, equity) economics can present us with a litany of examples of small-scale businesses that are commercially successful and, ecologically speaking, "do less harm." But doing less harm is not the same as sustainability or ecological permanence. Tasch's conclusion is not to disengage from capitalism and the growth imperative that has wrought so much destruction and for so long, but merely to put on the brakes a bit. This can certainly buy us some time while we shift to something entirely new and different, but the danger is that it reinforces the same practices and habits of thought—growth habit and idolatry of money—that are killing us. It also reinforces our slavish attitude toward capital which means that the bottom line of profit is always going to hold a higher position than the other goals of green business.

We need to listen to people like Jeanne and Dick Roy. We need to take a deep look inward and ask ourselves if we really believe such win-win promises, or whether these are just ways to make ourselves feel less guilty about what we are unleashing on future generations. The reality is that the win-win ideas of the triple bottom line and green capitalism have been tossed around for decades and throughout that time we have burned more fossil fuels, destroyed more topsoil, spewed more carbon, and engaged in more wars for oil. And as the saying goes, if we keep doing what we are doing, we are going to end up where we are headed. Half measures are not going to change that. Again the core challenge isn't just aspiring toward different values, though this needs to be done. The core challenge is our unwillingness to admit, or even consider, that the conventional, profit-driven models of business and our widely accepted expectations of financial growth and accumulation could be at the root of our ecological problems.

To be sure, projects of green or renewable technology can be showcased as both ecologically sound and profitable. But these are a very small segment of the otherwise unsustainable $15 trillion-dollar US economic system, or the $63 trillion global system. It is pure fiction to claim that this small segment will rise to carry the mantle of growth of this massive economic system that has always been driven by fossil fuels. Nonetheless, it's what people want to hear and if presented with enough fanfare, fiction can become reality in the popular imagination.

One way that green economics has captured popular imagination is by referring to it as "natural capitalism." Natural capitalism has a comforting ring of familiarity. It also suggests that if we are empowered with a bit of ecological know-how derived from nature, our economy can continue on with business as usual. Central to natural capitalism is the notion that production systems can be redesigned to emulate nature's innate efficiencies and recycling processes. Basic production processes can be revamped in a way similar to the regenerative capacity found in ecosystems. In the same way that a plant or animal dies and decomposes—which is nature's way of breaking down the parts and recycling them into new life—social systems of production can disassemble machines or finished goods as they wear out. The salvaged parts and materials can be re-used to produce more and new products. In this way, our processes of economic production can emulate nature's extraordinary capacity for efficiency and regeneration. And in a capitalist market economy, such close-to-nature efficiencies would be rewarded with profits and a growing market share.

This focus on efficiency raises a question. Efficiency is often referred to as a lighter ecological footprint. This means recycling and using resources in way that allows for economic production with minimal collateral damage. Efficiency gives us more output with less resource inputs and historically this has been a cornerstone of economic development. But there are hidden paradoxes with efficiency. When cars become more fuel efficient, the cost per mile goes down and, paradoxically, people drive more and fuel consumption does not decrease. Since 1980 the fuel efficiency of cars in the US has increased by 30 percent, but fuel consumption per vehicle has stayed about the same. And as the auto industry strives to sell more vehicles, fuel consumption rises. Similarly, as for-profit businesses become more cost efficient and their profit per unit goes up, they are compelled to produce more and grow. In the US, although the per unit amount of energy that goes into the production of GDP has decreased by nearly 50 percent in the last thirty-five years, total energy consumption has risen by over 40 percent.

Efficiencies gained from recycling will be important aspects of true environmental sustainability, but should not be equated with it. More often than not, though, this is what is done by advocates of green economics. Certainly a system that shifts to renewable energy, recycling, and more localized food production will be lighter on the footprint, but the scale and capacity for future growth of such systems is extremely limited. Any system that promises ongoing growth and expanding prosperity cannot be sustainable no matter how light our footprints may be. The US economic system has grown to a magnitude of $14 trillion on a foundation of cheap, one-time fossil fuels and

Americans burn through 21 million barrels of oil each day. It would be a wild stretch of imagination to think that we could substitute this magnitude of production and consumption with recycled materials.

In the discourse on limits to growth, the technology-will-solve-it arguments get raised with tiresome regularity. As I argue in all of my work, technology will be an important part of our efforts to achieve ecological permanence. But the technologies we develop to make us more efficient must be developed with specific goals in mind. If the goal is bottom-line profit maximization, then nothing is going to change except that we could possibly slow down the rate at which our system collapses. If the goal is genuine ecological permanence, then we can use technological efficiency to release labor time to develop the kinds of institutions and structures that will lead us down a path that is different from the one we are on now. Institutional change has to come before technological change. If technology remains conditioned by conventional business practices, then efficiency and productivity growth will only serve as means to profit maximization and market expansion. None of those is consistent with the broader aim of ecological permanence.

Another problem with the concept of natural capitalism is that it is grounded on the notion of the natural selection in market competition. Markets have always been used for trade and exchange. But conventional economic ideology sees a market system as an arena for the grim sorting process of natural selection—survival of the fittest as a law of nature. The companies that are most efficient as low-cost producers are selected to survive and high-cost producers are weeded out for extinction. Although there has always been brutal competition in capitalist systems, it has never actually worked this way. The fittest have never been those that are most efficient, but rather those that are most powerful. Also, capitalism as a system has never been about mere survival. It is about growth, accumulation, and expansion. Seen in this way, natural capitalism would actually be a dark and destructive way to emulate nature. Growth for growth's sake in nature is pathological—bacterial or viral epidemics, reproducing cancer cells, or rampant populations of invasive species. Biological systems have evolved for survival, not for eternal expansion. If they try, they quickly devour everything in their path and eventually lead to their own extinction by destroying their supporting habitats.

Brutal as all this sounds, Americans still have a close attachment to capitalism as if it were a law of nature. Most of us are ready to raise our glass to suggestions of having the marketplace sort out our problems. This is why policies aimed at fighting global warming tend to gain political support only when they are packaged as market-oriented. This would also explain why the only real politically acceptable way for Americans to deal with climate change is with proposals for true-cost pricing, carbon trading, and carbon offset programs in the open market.

True-cost pricing is based on a concept in economics called "externalities." Externalities are real costs generated in our production and consumption processes that are not reflected in market prices. When we clear cut forests in order to produce lumber for housing, the soil erosion and siltation of rivers that result are costly. Or when we burn gasoline to power our cars the damage caused by spewing carbon dioxide into the atmosphere is also costly. But the burden of these costs is mostly externalized—pushed off the producers' accounting books and onto a broader segment of society. Companies can appear to be more efficient by lowering their costs, but the costs are not actually lowered, just pushed into another sphere and hidden. The implication is that the prices we pay for things do not represent the full costs of production and they are much less expensive than they really should be. The artificial cheapness of things causes us to over-consume and do excessive environmental damage. Ultimately, however, the costs are felt in the form of suffering caused by all the damage we do to our life-supporting habitat.

True-cost pricing is making market prices reflect the real costs of production by internalizing those externalities. It is an attempt to get undistorted information about real costs into the markets. The assumption is that markets will work to lessen environmental damage if we can just get the prices right, because higher market prices that reflect true costs would slow down consumer impulses and thereby lessen the environmental damage we do. One way to achieve this, according to green economics, is to have market prices reflect true production costs with an excise tax of some kind such as a fossil tax imposed on fossil fuels.

Ideally a fossil tax would reflect the full cost of burning fossil fuels. The damage and costs generated by carbon dioxide effluents would be put back into the market price of fossil fuels and the burden be borne by those who burn them. Because of the extensive damage carbon emissions actually do, gasoline, diesel, and coal would all become much more expensive to buy. The companies that sell these fuels would have a harder time competing in energy markets as fuels that result in less carbon emissions would prove to be less expensive. By virtue of the natural selection of free markets, so the argument goes, fossil fuels will be naturally selected for extinction and clean energy will survive.

There is a huge practical problem with determining what the actual cost of the fossil tax should be. If Americans were to pay a true cost price per gallon for gasoline at the pump, what costs would that price reflect exactly? The cleanup costs of oil spills can be easily calculated in dollar terms. But what is the dollar value of the deaths and extinctions of species known and unknown that result from the spill? And what of the costs of global warming? Who is bold enough to claim that they know the dollar value of every creature destroyed or the costs of all the destruction wrought by melting ice caps, hurricanes, floods, drought, and wildfires? The costs of the damage stemming from global warming are systemic and incalculable and trying to calculate them would be as meaningless an exercise as trying to put a dollar value on life itself. Ultimately the line of demarcation between what we count and what we do not count would be arbitrary. Arbitrary measurements are unscientific and so there would be nothing *true* about true cost pricing.

Even if we ignore the tricky science there are still other intractable problems with this. True cost pricing with fossil taxes places the burden of costs on consumers. This is a kind of economic punishment inflicted on people for extravagant consumption habits. The presumed intention, therefore, would be to get them to change their habits. But recall that having extravagant consumers is precisely what our growth-driven economic system needs to keep going. People do not take well to being punished, particularly when they feel as if they have not done anything wrong. Consuming fossil fuels has always been the American way. It has been encouraged through advertising and marketing. Our economic system is built on mass production and mass consumption, and if people are consuming fossil fuels excessively, it is because our economic system needs them to. Like an autoimmune disorder, true cost pricing models amount to a system that is attacking itself internally and are nothing more than the cliché of attacking a symptom and not the disease.

Carbon trading and carbon offsets are other popular but failing strategies of trying to use the established market system to solve the problem of climate change. Carbon trading, also known as cap and trade, begins with a government plan for establishing a target maximum (cap) level of carbon emissions allowed for the year. Each year this target amount is to be reduced until the economy reaches an established goal. Climate scientists have concluded that the goal should be 350 parts per million (ppm) of carbon emissions by the year 2050 in order to avoid the worst effects of global warming. This would be about an 80 percent reduction from current levels. With its annual target established, the plan is to have the national government auction off pollution permits to companies that burn fossil fuels as their primary energy source—mainly utilities. The total number amount of carbon allowed with the permits would equal the targeted amount overall.

Some of these companies will be more efficient than others in how they manage their emissions. Those that innovate and discharge less than their permitted amount of carbon pollution will earn "carbon credits" and those credits can be sold to other companies that are less efficient and need to exceed their permitted amount. Theoretically this should make the innovative companies stronger than the ones that have to buy carbon credits to stay in business. The strong are rewarded and the weak punished in the natural selection of the open market.

For this to work, a carbon exchange must be established and the credits must be made into tradable commodities like wheat or gold. This is where investment banks come into the picture. Big Wall Street companies like Goldman Sachs and Morgan Stanley would create the markets and the tradable permits and earn brokerage fees in the deals. To the extent that they have been tried, cap and trade experiments have been exactly what one would expect from Wall Street-type institutions—fraught with greed, speculation, and mischief. Carbon permits can be worth big money, especially if they are subjected to a speculative bubble. Large corporations with lobbying power and influence pressure their governments into issuing free permits as a subsidy. As the permits are liquid commodities they can be sold in the bubble-inflated markets for a huge bonanza. Auditing carbon emissions is also an expensive and difficult process for the underfunded Environmental Protection Agency (EPA) to carry out. This makes it easier for companies to misrepresent the amount of carbon they discharge in order to get tradable credits that they can sell for cash as well.

Another problem in the cap and trade system is with "carbon offsets." Carbon offsets allow individual people or companies to earn tradable carbon credits by creating sinks such as tree plantations that will absorb more carbon out of the atmosphere to offset emissions. The carbon credits they earn can be sold in the open market for cash. For-profit companies are created as clearing houses to broker offset transactions. If a utility decides to sponsor a tree plantation project that technically absorbs more carbon than the utility emits, the company can earn credits to be sold and the clearing house company will broker the deal. Individuals who want to do something to fight global warming can donate money to these clearing house companies that oversee the plantations and keep the credits.

Typically these projects are done in developing countries as kinds of job-creation programs. Inspired by greed and not the desire to curb carbon emissions, companies have been known to raze forests, then plant new trees in order to earn the credits. Some businesses have been known to announce plans to expand their coal-burning operations by 200 percent, then later announce that they will cut their expansion back to 100 percent, claiming to have earned offset credits. As the offset business gets caught up in the speculative boom, businesses have an incentive to start plantations wherever they can. Often this means planting trees in places where the soil is healthy and there is water, which means a place that could be used for growing food. Rather than reducing carbon pollution, the markets actually offset food production and create food shortages.

Perhaps the worst thing about cap and trade and carbon offset policies is that they actually entrench the practice of burning fossil fuels. They were designed to be easy market solutions to global warming, but instead governments created a lucrative industry that is dependent on carbon emissions. If burning coal fades out, it so will the drive-out-the-carbon trading business, eliminating the fees earned by investment banks. Banks and government agencies become used to the fees and revenues from the trading industry and therefore have a vested interest in seeing that burning fossil fuels remains as a dependable practice. This has spurred environmental organizations around the world to fight against cap and trade. In 2004 some 300 environmental groups from around the world signed a document called the "Durban Declaration on Carbon Trading" to denounce carbon trading. The document declares,

> As representatives of people's movements and independent organizations, we reject the claim that carbon trading will halt the climate crisis. Carbon trading . . . turns the Earth's carbon-cycling capacity into property to be bought or sold in a global market. Through this process of creating a new commodity—carbon—the Earth's ability and capacity to support a climate conducive to life and human societies is now passing into the same corporate hands that are destroying the climate.[25]

Carbon trading and fossil taxes are again attempts to treat the symptoms and not the disease, and in fact, make the disease worse. Nonetheless these policies are fashionable because they are sold to the public with promises of win-win solutions. Promises of win-win presented in the conventional sense as Tasch suggests—fiduciary responsibility and care for natural ecosystems—are nothing more than red herrings. It is a fixation on making big money while trying to save the environment that distracts us from the real work of institutional change. And perhaps the biggest distraction of all is the outrageous claim that our $14 trillion economy can continue with business as usual and growth on a foundation of green energy.

Rethinking Green Energy

Renewable energy is the crown jewel of green economics. It is completely reasonable to contend that a sustainable future depends on harnessing clean and renewable energy sources. But if environmental economists claim that solar, wind, and biofuels are going to propel our $14 trillion-dollar growth machine into the future, they are truly living in the make-believe world of the alchemists. Americans sustain a daily diet of about 21 million barrels of oil, 2.7 million tons of coal, and 63 billion cubic feet of natural gas. Currently over 84 percent of our total energy use comes from these fossil fuels, another 8.5 percent is from nuclear, and only 7.3 percent comes from renewable sources. The main issue here is not whether or not renewable energy is efficient or cost effective, the issue is scale. It is possible to harness clean energy, but severe physical and financial bottlenecks will prevent us from using it on the scale we currently use fossil fuels. If we tried, the ecological and economic damage that would follow would be devastating: topsoil ruination, deforestation, water shortages, further depletion of just about all resources, and perhaps worst of all, food shortages.

One key difference between fossil and renewable energy is that fossil fuels are dense and ready-to-go sources and renewables are not. Renewable energy has to be produced from something else and that requires building massive and expensive infrastructure such as solar arrays, wind farms, new transmission systems, Stirling machines, algae ponds, biofuel conversion plants and the list goes on.

In the case of biofuels, it takes much land, water, and energy to get a relatively small amount of energy back. One of the greatest advantages of biofuels is that they have the potential to be carbon neutral. The vegetation used for biofuel feedstock is a carbon sink and can absorb carbon dioxide out of the atmosphere. If we don't overdo it, the sinks can reabsorb the same carbon emissions that get sent into atmosphere when burned. This could potentially be an ecologically closed loop process of carbon recycling and energy, but only as long as the rate at which the energy is burned does not exceed the rate at which it can be reabsorbed. The reabsorption process on the ground is slow because it relies on plant growth. Relying on biofuels in this sustainable way would require that we use far less than what we are used to with fossil fuels. Scaling down our energy to a sustainable level in this way is starkly inconsistent with the way our massive economic system functions.

To get a sense of this problem of scale, consider trying to replace gasoline with ethanol. Right now Americans consume about 140 billion gallons of gasoline every year. Gallon for gallon the energy density of ethanol is less than gasoline, so getting the same energy and work would take more, about 210 billion gallons of ethanol. The scale of this is beyond possibility as there simply is not enough land or water to produce that much.

The US domestic feedstock for ethanol is primarily corn. The most efficient ethanol processing plants are in the Midwest where they have a close proximity to cornfields. On average ethanol facilities can produce about 41.7 million gallons each year. To get 210 billion gallons of ethanol would require 5,000 new processing plants and would require growing 74 billion bushels of corn. This would be nearly six times the record corn output of 13.2 billion bushels. It would also require 368 million acres of the best farmland, which constitutes about 90 percent of the total cropland in the United States of which less than 13 percent is irrigated. Making ethanol out of corn requires an enormous amount of water. As it takes about 1,000 gallons of water to produce one gallon of ethanol, converting from gasoline to ethanol would take 210 trillion gallons of water. That would be about 4.5 times the total amount of surface and ground water currently used in agriculture. Ethanol also has a low energy rate of return, which means that we would burn almost the same amount of energy trying to produce it as what is gained in its production.

In some ways the industry statistics on corn-based ethanol are deceptive. Corn-based ethanol is a rapidly growing industry. In 2005, Americans torched 3.9 billion gallons of it and by 2008 the figure jumped over 9 billion. This growth could be interpreted as a success of the resilience of the market system's ability to adapt to changing circumstances. But this growth was not market-driven, rather it was driven by lavish federal government subsidies.

Given the obvious problem with corn-based ethanol, scientists and entrepreneurs have been looking for other alternatives. Recently algae has emerged as the next great green hope for biofuels. Algal biomass has a number of advantages over other feedstocks. One of them is that algae is a very resilient microorganism and can grow just about any place where there is water. It doesn't need to grow on arable land like corn, so land that could be used for food production does not need to be used to grow algae. Algae reproduces very quickly if it has the right amount of food and carbon dioxide. It naturally produces rich fatty substances, or lipids, that can be processed into biofuel. Another advantage is that algae can produce much more energy per acre than any other biofuel feedstock. Also, the kind of water algae grows in does not have to be clean or fresh, in fact it can grow in ponds of saltwater or water contaminated with toxic chemical compounds, raw sewage, or graywater. The promise of algae has brought in hundreds of millions in venture capital and research investment money such as a $600 million investment from Exxon Mobil in 2009 and hundreds of millions in government grants.

Like other biofuels, algae feedstock generates a very low energy rate of return. Algae needs much sun, large amounts of water, heavy concentrations of carbon dioxide, nutrients, and fertilizers. Growing algae in open ponds—the least energy intensive method—is like growing it in large swimming pools, but those ponds have to contain nitrogen, phosphorus, and carbon dioxide before the algae can grow into a viable energy feedstock. Unlike corn and other feedstocks that grow on land and can draw some nutrients and fertilizers from the soil, algae would rely almost entirely on external sources. That means chemicals made from oil and coal. One possible solution to the nutrient problem would be to build the algal ponds near urban areas so that you can use the nutrients in wastewater runoff to feed the ponds. The problem is that large amounts of energy would be required to pump the wastewater effluents into the ponds. The energy rate of return would be negative.

Getting enough carbon dioxide for the algae to grow is also a problem. Algae plants cannot absorb enough CO_2 out of the air to make biofuels and must rely on an external source to get the concentrations required to harness energy. The only practical way to get those levels of CO_2 would be to hook the algae ponds up to coal burning plants. Even so, by the time you burn the coal in a power plant, pump the CO_2 into an algae pond and account for dissipation, the amount of energy used up is greater than what is gained in the lipids. These problems aside, algae processing requires additional energy-intensive centrifuges to separate lipids from the pond water. Critics of algae biofuels argue that the proponents of algal biofuel are ignoring the laws of thermodynamics because the energy rate of return is negative—it takes more energy to produce algal biofuels than what is contained in the fuel itself.[26]

Even if those problems could be solved with technology, the main glitch again is scale. Just as with corn-based ethanol, scaling up algae biofuel production to match our current gasoline consumption would require a prohibitively large amount of infrastructure and land. To scale this properly, consider the largest and most efficient algal biofuel plant, which is located in Rio Hondo, Texas. The plant cost about $1 billion to construct, covers about 1,100 acres, and promises to produce about 4.4 million gallons of biofuels per year. The energy content is about the same as that from ethanol, but because of the low energy rate of return, which seems to be close to zero, there is little to be gained. Nonetheless, if we were to scale it up to our 140 billion gallons of gasoline equivalent per year, then we would need to build about 32,000 of those plants, covering 35 million acres of land. Not counting the land or the chemical nutrients, this would require a staggering capital investment of $32 trillion, well over twice the size of the entire US economy.

These renewable energy infrastructure costs themselves stand as prohibitive bottlenecks. Infrastructure bottlenecks are serious and intractable problems. Even the current oil industry is plagued with infrastructure bottleneck problems. In order to extract oil from four miles beneath the ocean or from tar sands, enormous amount of capital equipment is required. The more the industry shifts away from conventional oil production to these more extreme measures, the more material infrastructure—drilling rigs, pipe, transportation vessels, earth movers—will be required. Though oil companies are doing this, the amount of this infrastructure that can be developed is severely limited.

Developing large scale renewable energy infrastructure is even more limited. Coal-powered utilities in the United States use about a billion tons of coal to generate 2 trillion kilowatt hours (KWh), or 2 billion megawatts (MW), of electricity annually. If we tried to build up wind or solar infrastructure to reach this level of electricity generation, the costs would be out of this world. The first step in any economic development plan is capital formation. We have to raise financing first in order to build renewable energy infrastructure. Given the scale at which we use fossils, literally tens of trillions of new capital would have to be formed to build enough wind farms and solar arrays to match what we consume from coal. Where would this come from? Since renewable energy is only a small fraction of what powers the economy, it would have to be generated through conventional means.

Considering how much money the US currently spends on wars and bank bailouts, it may seem possible to build the infrastructure for some combination of green energy if given enough time. It is not unreasonable to assume that we could gradually build up a combination of biofuel, wind, and solar infrastructure while we are phasing out our use of oil and coal. While we are gradually wedging down the fossil fuels sector, we can be gradually wedging up the renewable sector. This will certainly happen to some extent, but the problem, again, is scale. If we tried to reach scale that would match fossils using renewables, we would become trapped in a much deeper problem—a financial bottleneck.

No matter how we look at this, the cost to finance these projects would be staggering. According to the International Energy Agency's *World Energy Outlook 2012*, a cumulative investment of $37 trillion in new energy infrastructure will be necessary between 2012 and 2035.[27] It is unlikely that nations could borrow that much from other creditor nations because there is not that much capital available to borrow as countries around the world are struggling with their own massive debt problems. The only way $37 trillion in new capital could be raised would be from a total accumulation of personal savings, retained business profits, and government tax surpluses, all of which would be generated through the regular business of producing and consuming things.

One problem currently is that the national savings rate for the US is negative. This means that not only are we not generating new capital, we are actually losing capital with each passing year and the difference has to be made up with debt. If this were to change and we returned to the robust 5 percent national savings rate of the booming 1990s, the question that would still remain is how fast would the economy have to grow and for how long before we accumulated $20 trillion in new capital. If we hold to the Social Security Administration's estimate of 2.2 percent annual growth and put all of the 5 percent national savings into a trust fund, it would take about twenty-three years of continuous growth and would add about $9 trillion to our current $15 trillion economy, and nearly all that growth would have to be powered by fossil fuels. We simply cannot do this. The principal reason we cannot is that we are passing the threshold of peak oil and the resource foundation that would support growth will be declining steadily throughout those 23 years.

This is the catch-22 that represents the absurdity of the win-win promises of green economics.

A recent report by economists at the International Panel on Climate Change (IPCC) projected a number of different scenarios for our energy future. The most optimistic is an estimate that renewable energy could potentially make up about 80 percent of the world's energy consumption by 2050.[28] Their forecast is based on two basic trajectories: the steady decline of fossils and the steady decline in the cost of renewables. The assumption is that as technology and innovation in renewable energy develop, costs will fall and renewable energy will become more cost effective. What is unclear in the report, however, is exactly how much energy overall will be consumed by mid-century. It might be possible to have renewables make up 80 percent of our energy consumption if we cut back that consumption to a quarter of what we use today. They also seem to be forgetting that renewable energy has a low energy rate of return and the infrastructure itself is built out of scarce resources. For these reasons it is unrealistic to assume that the costs will steadily decline for decades to come. But even with their most glowing optimism, the IPCC economists contend that to merely jump start something like a renewable energy revolution would cost somewhere between $3 and $12 trillion. The question therefore remains: how much damage would we do trying to grow our economies fast enough to raise that much capital?

The long-term solution is the one that the economic alchemists and people in general don't want to hear. To achieve true ecological permanence, humans will have to get by with much, much less production and consumption. There is no question that we can continue building solar arrays, wind farms, bio-fuel conversion plants, carbon scrubbers, hybrid or electric vehicles, and all the other accoutrements of a green economy. This work can generate good paying jobs as promised and can make returns to investments that will provide capital for further growth. But this can only be done on a very small scale and only for those who already have a surplus of capital to draw from. Otherwise, with a shrinking resource base, any new economic growth would be zero sum growth, which will create more prosperity for some and worsen the conditions of poverty for others. These so-called "others" are not only people in some distant land where they can be easily ignored. They are in our own communities, and unless we want a permanent state of conflict and warfare in our own communities, we will have to begin rethinking what it really means to create a green economy.

Green energy will indeed be a central part of our sustainable future. In areas where there is abundance of sunlight, then solar energy will be more effectively developed. Solar energy makes less sense in the cloudy and soggy Pacific Northwest, but hydroelectric power does. The Midwest may rely on wind and biofuels to some measure and so on. Given the geographic circumstances, these energy sources will naturally develop. The question raised here is *how much* of that energy will we realistically be able to consume while achieving stability and ecological permanence. In other words, how far down would we have to scale our levels of production and consumption to achieve ecological permanence and be nestled within the carrying capacity of the planet? This is a very difficult question to answer because we don't know what technology will bring in the future. But at present, the scientists at the Swiss Federal Institute of Technology estimate what they call a "2,000 Watt Society." This means that on a planet that supports over 6 billion people where every person consumed an equal amount of energy, the amount of energy consumption would be the equivalent of about 2,000 watts per day. The more people we add to the planet, the lower the daily diet of energy would be. Currently, Americans consume about four times that amount. This 2,000 watt benchmark might be seen as a goal that would achieve energy equality and would avoid resource wars or energy genocide as wealthy countries starve poor populations to death by depriving them of energy. But to get there we would have to scale down our energy consumption by 75 percent. What would that mean? It would mean that our social system of production would have to evolve into something that uses drastically less energy and that means a completely different system.

This evolutionary process is crucial because the systemic trauma of economic contraction to such a large degree would be catastrophic. To quote Ted Trainer, author of *Renewable Energy Cannot Sustain a Consumer Society*, 2010:

> There is a widespread assumption that a consumer-capitalist society, based on the determination to increase production, sales, trade, investment, 'living standards,' and the GDP as fast as possible and indefinitely, can be run on renewable energy ... But if this assumption is wrong we are in for catastrophic problems in the very near future and we should be exploring radical social alternatives urgently.[29]

Green economists that contend it is possible to power our consumer society on renewables are giving people a false sense of what is possible. For a century we have powered our massive economic machine on oil, particularly the second half. This is also combined with a heavy use of fossil water and other resources that we cannot replace with new technology any more than we can recreate the sun. If we are to scale down to a level consistent with the carrying capacity of our planet, we will have to evolve new institutions that will allow us to focus our economic activity closer to home and do it on a much smaller scale. This should be the primary focus of our great transformation.

To some degree, transition communities must become shadow communities and break away from nearly all our conventional habits and expectations. The transition has made adapting to energy descent one of its primary goals. Every community will have to begin forming an energy descent action plan and prepare to become more self-reliant and we will have to do it in a way that is fair and stable. Community-based planning and action means making a fundamental transition in the locus of economic activity from a global scale to local. Jeff Rubin argues that this is why all of our worlds are going to get smaller. But simply emphasizing economic localization by itself is not a solution. Local institutions have to be made anew and cut from very different cloth.

Rethinking Localization—Small Is Not Always Beautiful

There is a clear strong connection between resource limitations and economic localization, or re-localization. Proximity matters and it will matter even more in the years to come. The scarcity and costliness of energy will render obsolete the current global economy that is built on a vast international division of labor. Globalization will eventually be replaced by localization—for better or worse. Aside from bumper sticker sentiments, going local is not automatically better for communities; it has to be consciously made so by an active and informed citizenry.

The inevitability of localization is not exactly news to many of us who have been watching this process. One organization in particular, The Business Alliance for Local Living Economies (BALLE), has been at the forefront of community based efforts to go local. The organizers of BALLE have been committed to building a network of small business that would constitute a business-oriented grassroots movement for overall economic change to a more community-based model that would stand in opposition to corporate globalization. BALLE was organized in 2001 as a result of a conference dedicated to forming local, socially responsible business networks. The organization was formed on the principle that communities that foster the development of local business ownership tend to be more prosperous in the sense that they create more local wealth, local income, local tax revenue, and local jobs.

Though BALLE opposes Wall Street and giant corporations and favors local businesses, the 20,000 businesses in its membership are quite conventional. Their organizational rhetoric is the same as that of the green economists who claim that going local and going green can also be handsomely profitable. One of the intellectual leaders of BALLE, Michael Schuman, waxes ebullient about how the small business sector creates more jobs and earns more profits than Fortune 500 corporations. This is a good thing as long as we are holding on to conventional measures of success. The emphasis on "Living" in the Business Alliance for Local Living Economies suggests some commitment to ecological permanence, yet member businesses are not held to any rigorous standards for sustainability.

There is no reason to think that just because businesses are small they are going to be any more community-oriented or ecologically sound than large businesses. As they are driven to maximize profits just as much as large corporations, small businesses adhere to the same conventional standards for success—bottom line profitability and growth. Looking at their historical record, local businesses in the 19th and 20th centuries were more likely than not to be slave plantations, quasi-feudal company towns, neighborhood loan sharks, or Dickensian sweatshops. Many small businesses today maintain dubious track records on social and environmental well-being. They often lobby against local zoning laws designed to protect watersheds and green spaces or for more self-reliance in food production. Many small farmers, often eager to subdivide their land and sell it off in parcels to developers and big box stores, lobby hard against land use planning and urban growth boundaries that are important responses to rising energy costs and fuel scarcity.

The sad truth is that small businesses in America are less likely than large ones to pay for health insurance for their employees. Through their lobbying organization, the National Federation of Independent Businesses, small businesses lobbied hard against mandatory health care legislation in both 1993 and 2009. As part of their individual claim to property rights, small businesses also lobby hard against health and safety inspections; they also lead the local fight against property taxes, minimum wage or living wage ordinances. Doug Henwood reports in his newsletter, *Left Business Observer*, 2010, that the notion that virtue and size are inversely related is a populist fantasy. Commenting on Arianna Huffington's Move Your Money campaign against the large commercial banks, he notes that the banks recommended on the movement's website for his neighborhood in New York finance gentrification of black neighborhoods are vested in Treasury bonds rather community-based assets, and finance real estate development projects that will squeeze out rent-controlled units in favor of posh, unregulated apartments.[30]

If we are planning for an ecologically permanent and fair local economy, we have to create new business models that will foster a different set of values. Small and local businesses that only consider their bottom lines will be of no help. E. F. Schumacher claimed that "small is beautiful" but small is not necessarily beautiful, we have to make it so.

Conclusion

The common failure of win-win green economics and localization is that there has been no serious attempt to design social enterprises as real alternatives. Our economy and culture are plagued by attempting to solve problems with the same habits and conventions that created them in the first place. Trying to use markets and for-growth models of business to deal with ecological problems seems amazingly naïve and nonsensical. The imperative for real change is looming large and we have to begin the process of developing truly alternative economic models and institutions.

To do this, we need to become visionaries. We must cultivate a vision that can extend beyond the borders and contours of our current system, and to see through the fog of mythology that helps sustain that system. At the risk of sounding cynical, win-win proposals usually require a boatload of mythological justification. Without such alternative vision, we can only see what is presented in terms of conventional models of growth and consumerism, and we remain blind to the possibilities of real change. We also become susceptible to pseudo-solutions that are packaged in hype and marketing strategies and cannot lead to anything truly different because they are born out of the same mindset and practices that are bringing us to the edge of ruin. Without this vision our ideas will remain stale and contradictory, and our practices, even with all of the best of intentions, will lead only to more trouble. We remain trapped in a maze that leads only to dead ends. A truly different vision can help guide us to create truly alternative economic institutions, but we have to find a way out of the maze to do it. If we succeed, we have the potential to contribute to a more graceful evolution of our culture.

Institutional development is itself a kind of technological development, though it is not typically seen as such. Mostly technology is viewed as something engineers create and is nearly always centered on something mechanical or electronic. Rarely are technological innovations viewed critically. It seems to be a forgone conclusion that new technology will always be better in the sense that computers and word processing programs are better than typewriters. And as long as you can make that familiar analogy with plausibility, only the narrowest curmudgeon would be reluctant to nod in pious agreement. But this presents us with an interesting irony.

When most people think of technology curmudgeons, they think of someone of advancing years who appears angered and intimidated by computers, the internet, or wireless technology that they don't understand. This has been a familiar and condescending image for decades. Fear of being branded as such a curmudgeon has created a certain shyness or reluctance to speak out on the shortcomings of the so-called technological fixes. But the condescension itself has created a new class of curmudgeons—those who are angered and intimidated by these critics.

People attached to business as usual are reluctant to face the scientific realities that scarce energy and limits to growth have created. To face this would inevitably lead to the conclusion that our culture and institutions must change. The new curmudgeons are frightened and alienated by this. The narrow-minded majority in the 19th century accepted the institution of slavery with clichés about genetic differences. And today there are droves of people who refuse to face the realities of energy descent or climate change by taking comfort in saying things like, "Well, at least it won't happen in our lifetime" or "Technology will come through." There is always some short-term consolation derived from deluding ourselves, and there are always long-term consequences.

When Thomas Malthus argued that human population growth would outstrip the world's food supply, critics derided him for his failure to take technology into consideration. When the Club of Rome first published its report on limits to growth, their suggestions for solutions were largely technological, but resource depletion accelerated nonetheless. When the world leaders gathered at Rio de Janeiro for their Earth Summit, they also called for technological solutions and resource depletion continued to accelerate. It seems clear now that any efforts to deal with these problems must be grounded in some measure of genuine institutional change.

CHAPTER FOUR

E. F. SCHUMACHER AND THE META-ECONOMISTS

Claims that we can make our growth-driven capitalist system ecologically permanent are contradictory and confusing. Though there seems to be a general assumption among the green economists that we have to change our production and consumption habits, when they get down to the specifics they tend to cling to established conventions. Recall that Woody Tasch suggests we should listen to E.F Schumacher and others who hold economics accountable to higher morals and aesthetics, but then backs away from his own suggestion because those standards tend to clash with corporate interests. In doing so, he is conceding that corporate interests precede all else, even our most heartfelt convictions about what is good, beautiful, or wholesome. Instead he and the other apostles of green capitalism offer half measures or pseudo-solutions to ecological and social problems. This circumspection amounts to self-censorship and is perhaps the most effective mechanism for perpetuating the very habits that are responsible for the ecological and social problems in the first place.

We need to turn Tasch's approach upside down. Instead of subordinating our moral and ethical standards to the business of profit-making, we should be cultivating our heartfelt convictions about what is good, beautiful and wholesome and see to it that the business of producing and consuming things is supporting us in a way that allows us to live according to those convictions.

Those innermost convictions are what economist and philosopher E. F. Schumacher calls meta-economics. Meta-economics is a core belief system that transcends economics and other disciplines. It is a metaphysical center that cannot be proven or disproven by empirical science. Rather it is what we believe unconditionally to be true and is what guides our judgment and reasoning—for better or worse. Schumacher lamented that ". . . it is our central convictions that are in disorder, and, as long as the present anti-metaphysical temper persists, the disorder will grow worse."[31] In other words, the ecological and social problems we face are symptomatic of a much deeper defilement of our core beliefs. His point was that contemporary life is contaminating our core beliefs with cynicism and materialism such that, "our heart and mind are at war. Our reason has become clouded by an extraordinary, blind, and unreasonable faith in a set of fantastic and life-destroying ideas inherited from the 19th century."[32]

Schumacher and the other kindred spirits I will cover in this chapter argue that we can free ourselves from these life-destroying ideas and cultivate new ways of seeing the world around us. We can create healthier ways of being that involve changing how we meet our basic needs, cultivating a spiritual life, developing a deeper sense of aesthetics and the meaning of good work, realizing a sense of true justice and personal empowerment, and fostering a sense of belonging in our communities. As early 20th century intellectual Randolph Bourne put it, to create a "beloved community is the cultivation of that ripening love of surroundings that gives quality to a place, and quality, too, to the individual life."[33] I call them the "meta-economists" because they hold these convictions above the principles of economic production and consumption.

The meta-economists explored here are Anglo-American visionaries. As their ideas were unconventional in their time and clashed with orthodox belief systems, they were frequently met with scorn and hostility. Yet these intellectuals were undaunted and established a legacy of original thought. From John Ruskin and William Morris in the mid-19th century to Lewis Mumford and E. F. Schumacher in the 20th century, a common theme in their work is that they were all prompted to breathe new life into communities as an antidote to dreadful modern industrial systems. The industrial revolution did bring great material improvements in people's living standards. But these visionaries saw that it also created conditions of dehumanization, environmental ruination, and the poisoning of the human spirit.

Another common thread among them is their commitment to both an inward and outward transformative experience. That is, they saw that as individuals strive to better the conditions in their communities, they improve their own psychological and spiritual well-being; and as they evolve inwardly, they are naturally able to improve the conditions out in their communities. These meta-economists implore us to go out and engage in civic renewal, and by doing so we return transformed psychologically and spiritually. They make no distinctions between aesthetic, spiritual, intellectual, and political transformation, for they are all inseparable facets of human existence. The defilement of one facet leads to the destruction of the others; care of one facet leads to the well-being of the others.

The end of the Oil Age is arriving and it will be necessary for us to embark on our own heroic journeys. One way or another, our lives are going to change. The question is whether we can guide this change toward something better. We can find inspiration from these visionaries whose ideas are as relevant today as they were in the 19th and 20th centuries when they were creating them; perhaps more relevant. They envision people making the hero's journey to democratically re-create community-based economies and cultures in ways that restore meaning and joy to work by revitalizing craft traditions; and they envision these local economies to be fair, stable, and technologically appropriate, with a commitment to preserving the beauty of nature. Perhaps if we take these ideas to heart as we make our great transformation to a world without fossil fuels, we'll be able to do so with grace and with a deeper sense of well-being.

The Early British Meta-Economists

This legacy of meta-economists and visionaries begins with John Ruskin. Ruskin was a mid-19th century British art critic and writer who covered art and architecture, culture, and economics. He was also a social reformer and sought justice for working people, though his sense of justice was unlike that of other social reformers of that period as it was tied to artistic expression rather than party politics. He believed artistic expression is also bound to spiritual well-being. Work, aesthetics, and spiritual fulfillment are, for Ruskin, one and the same. He saw that humans can rectify their suffering by looking inward to see both their ignorance and their inherent capabilities. If people overcome ignorance, cultivate their capabilities, and bring those capabilities into society, society will become a better place to live. But remaining in a state of ignorance thwarts personal development, and as the individual remains stunted, so does society. For Ruskin, fully developing our inherent abilities is to foster creativity and expressiveness in work. Work is necessary for life, but it is not only the fruits of our labor that give us joy. Ruskin saw that there is joy in the work itself as we have the innate capability to do it with creativity and artistic sensibilities.

Ruskin's early work was mostly preoccupied with art and architecture. He felt that great art must be done with a sense of joy, but that sense of joy is as much social as it is personal. The artist must have a sense of personal freedom as well as a sense of belonging to a community and being appreciated by the community. Ruskin also held steadfast to the belief that authentic artistic expression must convey a deep understanding of truth and nature. Truth is derived from well-developed human capacities and not from learning in the conventional sense. The artist or artisan must have the mastery of his medium, and his ability to convey truth is inseparable from a moral and spiritual outlook. That is, for Ruskin, genuine art is not created from "taste" as it was seen in traditional highbrow culture, rather it involves the entire moral, spiritual, and creative person. For both the artist who creates the work and the person who perceives it, true art involves emotion, intellect, morals, spirituality, and other human capacities.

In Ruskin's view, great art must also come from an understanding of natural, organic forms. He contended that all organic forms, including human beings, are at the height of their potential beauty when they are allowed to develop perfectly according to their innate laws of growth. He saw organic forms existing in a state of interdependence. A plant living in nature requires the help of insects, birds, and other plants for it to become wholly developed. If any one of these elements is removed or destroyed, the plant will be stunted in its development and will not be able to reach its full potential for beauty.

Ruskin criticized the fashions of modern art for their failure to portray such organic interconnectedness. Modern art in his time was dominated by a movement to revive classical artistic forms of the late Renaissance. Ruskin rejected classical forms as reductionistic, standardized, mechanistic and devoid of any grounding in nature or truth. Ruskin leveled a scathing critique against modern architecture as:

> . . . unholy in its revival, paralysed in its old age . . . an architecture invented, as it seems, to make plagiarists of its architects, slaves of its workmen . . . an architecture in which intellect is idle, invention impossible, but in which luxury is gratified and all insolence fortified.[34]

This sociological view of art was typical of Ruskin. He saw the art of the 19th century as a reflection of deeper social and aesthetic maladies stemming from *laissez faire* capitalism and the industrial revolution. For Ruskin, capitalism did not represent a system of free markets and liberty in economic activity. He viewed it as an oppressively mechanized and dehumanizing system designed to benefit only a small few. Ruskin saw the classical aesthetic of art in his time as contrived as the classical notion of "economic man" who is portrayed as a passive, self-interested blob of utilitarian desire without higher capacities for individual development.

By contrast Ruskin celebrated Gothic art and architecture. For Ruskin, Gothic architecture with its tall spires and gargoyles represented organic relationships between craft workers and their guilds, the workers and their local community, the community and nature, and between people and God. Gothic forms represented for Ruskin "the liberty of every workman who struck the stone."[35] For Ruskin, true Gothic revival and the free expression of craft workers would be an antidote to the repressive sweatshops of industrial capitalism. Gothic revivalism for Ruskin would be a new movement that integrated natural aesthetic forms and human labor. As such, the meaningfulness of productive work would be restored to the stonemason, the carpenter, and other skilled workers.

He rejected capitalism for other reasons beyond the oppressive working conditions of the factories. Like Karl Marx and many of his followers, Ruskin viewed capitalism as a system that created great wealth at the expense of the poor. He also saw human potential ravaged by competitive markets. A market environment in which an individual worker is pitted against every other worker for wages would result in the destruction of craft traditions in a kind of race to the bottom. Wages would be bid down to levels of desperation at which the proper skills and techniques for the craft would be impossible to maintain. Working people, as they were deskilled with mechanization, would lose their ability to be creative and would therefore become stunted in their personal development and stripped of their role as the stewards of craft labor.

Unlike the revolutionary socialists, however, Ruskin did not see the future of the workman as taking over the factory system in revolution. Rather he envisioned creating wholly new alternatives to the factory and market system. He conceded that doing away with industrial capitalism would require some level of state socialism, but more importantly, he saw it as being replaced by a certain social economy. The social economy he envisioned is a network of social institutions created for the well-being of the worker and the community such as public institutions, community-based organizations, worker cooperatives, and guild-like craft organizations.

Ruskin's ideas met with some hostility by those seeking to uphold the way of industrial capitalism and classical architecture, but his ideas left a strong impression on many who were open to social change. One person in particular was another meta-economist, William Morris.

William Morris was a 19th century British artist, art critic, and socialist. Morris shared Ruskin's disdain for modern architecture and extended the same criticism to the industrial production of decorative arts. He also shared Ruskin's interest in Gothic revivalism. Morris was enormously creative in his own work which included handcrafted decorative objects, Gothic style printing, upholstery fabric, furniture, stained glass, poetry, and architectural design. He championed artisans who produce handicrafts as true artists in their own right, claiming that there should be no hierarchical distinctions in artistic mediums.

Like Ruskin, Morris saw that artisan work and politics were inseparable. Both had deplored the conditions within which working people toiled in 19th century England, and saw craft labor and the empowerment of creative artisans as the solution. Both were advocates of traditional craftsmanship using natural medieval and folk decorative patterns. This blend of social reform politics and craft tradition based on natural forms led Morris to instigate what was succinctly named the Arts and Crafts Movement.

The Arts and Crafts Movement was an attempt to revive a sense of community centered on artisanship. For Morris the movement was keyed to the abolition of the dehumanizing industrial division of labor that English poet and artist William Blake referred to as the "Satanic Mills." Morris advocated establishing a more localized, cooperative system centered on an artisanal mode of production. He proclaimed that, "one day we shall win back art, that is to say, the pleasure of life; win back art again to our daily labor." For Morris, artisan labor "used the whole of a man for the production of a piece of goods, not small portions of many men."

For Morris, establishing a social economy based on artisanal labor would require very different institutions for organizing labor than the market system. Morris envisioned a class of craftsmen/citizens leading the social organization of productive work. Though he had been involved in socialist politics, like Ruskin he did not share the goal of the Fabians or other socialist groups to take over the factory systems for themselves. Morris hoped that working people could form associations of producers in skilled craft traditions in which the workers were able to reclaim their skills and heritage that had been lost to the mechanized wilderness of capitalist factories. Morris saw these associations, led by these skilled artisans, evolving into a broader system of production or commonwealth.

Taken together the work of Ruskin and Morris stands as a kind of polemic against modern art, orthodox economic theory, and revolutionary socialist ideology. Both the modern art and economic theory were seen by them as a rationalization of a dehumanizing, mechanized system that was devoid of any sense of nature, beauty, or the creative human spirit. Orthodox economics treats human labor as inherently irksome and nothing more than means to ends. By contrast, Ruskin and Morris saw it as a necessary part of a joyful life. In this sense, Morris and Ruskin were meta-economists. Their work comes from a spiritual center intertwined with a sense of aesthetics, a reverence for the beauty of nature, a deep sense of compassion for working people, and heartfelt conviction about the joys of creative work.

Patrick Geddes and Ebenezer Howard were among the next generation of visionaries who were influenced the by Gothic and aesthetic revivalism of Ruskin and Morris. They upheld the value of aesthetic form, artistic creativity, spirituality, and the love of nature. But rather than focusing on social reform movements through the revival of artisanal traditions, their efforts were focused on changing the abysmal conditions of the industrialized cities. They sought for a great transformation in urban landscapes.

Patrick Geddes was a late 19th century biologist from Scotland. As a biologist he imbued Ruskin's Gothic revivalism with a kind of organic and ecological vision. Like Ruskin and Morris, he held on to a romantic view of organic forms, and his view of biological evolution differed from that of Charles Darwin and Herbert Spencer. Darwin and Spencer viewed biological evolution through lenses colored by capitalistic notions of individualistic competitive struggles for survival. The idea that species evolve as a result of individualistic struggles for survival was what Schumacher lambasted as one of the "life-destroying ideas inherited from the 19th century."

Geddes saw evolution progressing not from natural selection or survival of the fittest among individuals, but as a result of whole species forming symbiotic relationships and wholesome interaction with their habitats. He viewed forming cooperative biological relationships as more significant than competitive contests for survival. For Geddes this applies to humans as well. Humans are not so much "rugged individualists" as they are dependent on functioning within a vast web of biological and social relationships. In other words, one of Geddes's core convictions was that humans are more social and ecological than individualist beings.

It was from this conviction that Geddes sought to start a movement in urban planning centered on realigning urban living spaces in an ecological way. Like Ruskin and Morris, Geddes was appalled by urban conditions dominated by industrial machine systems. Life could be improved by recreating our living spaces to nestle harmoniously within more natural and organic systems where human culture forms into a living ecology of natural relationships. His ecological and interdependent view of human existence starkly contrasted with the beliefs of the classical economists who saw ideal human behavior as atomized, self-interested, and doomed to endless combat in open markets.

Geddes chose to view natural systems in a more cooperative existence rather than one characterized by brutish competition. He felt that it would be better not to assume our existence is keyed to exploiting nature to gain competitive advantage, but to cohabitate with, and be the stewards of, nature. For Geddes, compassion, sympathy for people and the environment, and cooperation would create synergy—an emerging condition that would be far more beneficial to everyone than if each were to pursue their own individualistic agenda. His was a prototypical vision for an urban permaculture community based on mutual cooperation with each other and nature such that all organic forms are allowed to develop to their full potential.

Geddes envisioned that the whole person—a fully developed biological and social being—would also participate in urban community life. A person not stunted by the industrial division of labor could become fully engaged in transforming the conditions of urban life. Inspired by Ruskin's aesthetic revival, this life could evolve into a kind of new "industrial reformation" in which there would be a re-organization of production systems that unite art and science to form a mode of production that is aesthetic, ecological, and productive. In other words, Geddes combined Ruskin's moral and aesthetic criticism of industrial capitalism with insights gained from biological science to form an aesthetical-ecological-economical view of production systems.

With this holistic view, Geddes created a *eutopian* vision of planning. As opposed to utopia which is a vision of something that does not exist, Geddes's eutopia is something that exists when the conditions are just right. A eutopian human existence would be one in which the social and environmental conditions are perfect for contented living. With this vision, Geddes developed a template that would link social reform movements, city planning, and a harmonious relationship with the natural environment. According to Geddes, the goal of these movements is what he called "civic regeneration." He saw civic regeneration as reconciliation between antagonistic aspects of life in industrialized cities where ideals of how to live healthy urban lives are in conflict with the practical concerns of commerce, where people's lives in the country are strikingly different from lives in the city, and where aesthetic forms of expression clash with ugly manifestations of heavy industry.

For Geddes, as for Ruskin and Morris, the task at hand is a transformation. That is, a cultural renewal in which modern societies build an entirely new way of life and culture that will enrich the individual person, integrate nature and aesthetics, and restore the bonds of association that will give rise to a healthy socio-ecological community. To achieve this, Geddes argued, we must maintain an undying faith in living organisms, including humans, to respond creatively to their environment. This faith in the potential for life to remake and revitalize the surrounding material world was the guiding principle behind Geddes's concept of civic regeneration. For Geddes, organic life's ability to revitalize itself was a kind of insurgency or subversive force for change that communities will have to tap into for their renewal and resilience.

Like that of so many other visionaries, Geddes's work was largely underappreciated in his time. Nonetheless, as we will see, the early 20th century thinker, Lewis Mumford, found an intellectual "father" in the work of Geddes. His ideas resonated with other like-minded visionaries around the turn of the 20th century such as anarchist Pyotr Kropotkin, institutional economist Thorstein Veblen, and urban planner Ebenezer Howard. Common to all of them is a vision for creating an economic transformation that shifts away from the ravages of the industrial system to a decentralized economy of small producers and cooperatives in communities that balance urban culture with the aesthetic and healthful benefits of nature; and that upholds the virtues of craft and local citizenship against the parasitic forces of capitalism.

Ebenezer Howard was also a visionary whose work was met with hostility during his time, yet eventually proved to be enormously influential. His vision was to create "garden cities" for the future. Like Geddes he wanted to see an end to the artificial separation between city life and nature, and was disgusted with the slum conditions of modern cities, particularly London where he spent most of his life.

As a young man HowardΩ left London for Nebraska in order to try his hand at becoming a farmer. After realizing that he lacked the acumen for farming, he returned to the city—first to Chicago and then back to London. While in America, he became fascinated with the transcendental poetry of Ralph Waldo Emerson, Henry Thoreau, and Walt Whitman, and was particularly moved by Thoreau's vision of simpler ways of living with nature as a model for social and cultural conditions, though his experiences of both urban and rural living threw him into a state of ambivalence toward both city and country lifestyles. He sought to catalog the differences and find a way to combine the best of both worlds into a single project that would provide a rough blueprint for living spaces of the future. His work culminated in a book originally titled, *To-morrow: A Peaceful Path to Real Reform* (1898) and subsequently reprinted as *Garden Cities of To-morrow* (1902).

In *Garden Cities*, Howard constructed a three-section diagram called, "The Three Magnets." In one section, or magnet, called *town*, he outlined both the positive and negative aspects of city living. In Howard's view, cities had much to offer and attracted people to their vibrant cultural centers, modern technology, sophistication, amusement, and economic opportunity. Yet cities also were characterized by overcrowding, slums and stark wealth inequalities, high cost of living, long distances between workplaces and people's homes, pollution, long work hours, and water shortages. Urban living was a kind of tragi-comedy for Howard, as so many people were drawn to the glitter and amenities of urban living, only to suffer in the squalor of poverty-ridden slums.

The second magnet in his diagram, *country*, is the list of pluses and minuses of rural life. Rural living lacked "society" or a certain vibrancy; people had little to do and opportunities for employment were scarce. He was disturbed by the feral attitude toward property reflected in notion of "trespassers beware." The countryside was not developed adequately and lacked adequate sewage facilities. Howard witnessed economically depressed and deserted rural villages and a general lack of public-spiritedness and amusement. Yet he admired the scenic beauty of the forests and meadows, fresh air, sunshine, and the notable benefit of affordable housing. Rural life had scenic beauty, but the effect on the mind and spirit could be dullness and backwardness.

The logical course of action for Howard was to design entirely new living spaces that were based on a combination of positive aspects of both town and country. This hybrid was the third magnet, which Howard called *town-country*. The town-country magnet was a model that would combine the vibrancy and opportunities of the city with certain scenic and healthful qualities of rural living. This model formed the basis for his garden cities of the future.

In addition to the transcendental poets, Howard was also inspired by John Ruskin's notion of combining an aesthetic sensibility with the daily work of economic activity. In *Garden Cities of To-morrow*, he opens his chapter on *town-country* with a passage from John Ruskin's small book, *Sesame and Lilies* (1865):

> Thorough sanitary and remedial action in the houses that we have . . . so that there be no festering and wretched suburb anywhere, but clean and busy street within and the open country without, with a belt of beautiful garden and orchard round the walls, so that any part of the city perfectly fresh air and grass and sight of far horizon might be reachable in a few minutes' walk.[36]

Following along with Ruskin's aesthetic view of living space, Howard begins by imploring his readers to use their imagination of a utopian experiment. The experiment is the garden city, the town-country magnet that is designed to capture both the vibrancy and opportunities of the city as well as the scenic and healthful qualities of rural life. It is a relatively small-scale, self-sufficient town that is inspired by the aesthetic sensibility expressed in Ruskin's *Sesame and Lilies*.

Howard was just as much compelled by a vision for building new institutions as by his sense of aesthetics. He outlined an entire city project that would begin with the acquisition of a 6,000 acre section of agricultural land financed with a 4 percent interest-bearing mortgage bond. The institutions that comprise the garden city are created for the specific purpose of creating new industries, jobs, small businesses, cooperatives, and healthy surroundings on land administered by a municipal central council. Howard writes,

> Its object is, in short, to raise the standard of health and comfort of all true workers of whatever grade—the means by which these objects are to be achieved being a healthy, natural, and economic combination of town and country life—and this on land owned by a municipality.[37]

Howard's vision was not just a model for city planning, it was a sociological model—a model for another way of life. His utopia would combine aesthetic beauty of the park with the cultural amenities of music, literature, art, and science and all of these would be placed at the core of the garden city. People's homes were to be affordable cottages situated between the scenic and cultural centers on one side and their places of employment on the other. Houses and workshops built on the land were to pay rent to the municipality in order to pay off the bond mortgage and to provide resources for public works: infrastructures, civic buildings, schools, and parks. Howard closed his chapter on town-country in the original 1898 edition with another quotation from Ruskin, *Unto This Last* (1860):

> No scene is continuously and untiringly loved, but one rich by joyful human labor; smooth in field; fair in garden; full in orchard; trim, sweet and frequent in homestead; ringing with voices of vivid existence. No air is sweet that is silent; it is only sweet when full of low currents of undersound—triplets of birds, and murmur and chirp of insects, and deep-toned words of men, and wayward trebles of childhood. As the art of life is learned, it will be found at least that all lovely things are necessary;—the wildfire by the wayside, as well as the tended corn; and the wildbirds and creatures of the forest, as well as the tended cattle; because man doth not live by bread only, but also by the desert manna; by very wonderous word and unknowable work of God.[38]

Like Ruskin, Morris, and Geddes, Howard was derided for his idealism. Yet he was undaunted by his critics and went on to launch a garden city movement. He founded the Garden City Association in 1899 that eventually evolved into the Town and Country Planning Association, which remains today as Britain's oldest environmental organization. The association planned to create a prototype village based on Howard's model. The village chosen was Letchworth in Hertfordshire north of London. In 1903, Letchworth became the world's first garden city.

The garden city idea gained momentum and other models were developed in nearby Welwyn, also in Hertfordshire, as well as in Germany, Latvia and as far as away as Australia. To honor his efforts and legacy, Howard was knighted in 1927, one year before his death. His ideas lived on to influence generations of urban planners worldwide, including Walt Disney's vision for a planned city known as the Experimental Prototype Community for Tomorrow (EPCOT), now the site of the famous EPCOT Center. And in 2007, architectural writer Jonathan Meades tipped his hat to Howard in his "Jonathan Meades: Abroad

Again" travel series *Heaven: Folkwoven in England*, "[The Garden City was] a social experiment on par with the Welfare State, a social experiment that affected us all and still does." Among the many affected as such was Lewis Mumford who wrote, "Nothing could be a more timely contribution to building a life-centered civilization than the publication of Sir Ebeneezer Howard's famous book."[39]

The Young Americans

In the first few decades of 20th century America, cultural critics Randolph Bourne, Van Wyck Brooks, Waldo Frank, and Lewis Mumford continued the tradition established by the British meta-economists. Together they comprise what cultural historian Casey Nelson Blake refers to as "The Young Americans." Their work stands alongside that of Dewey, Veblen, Emerson, Whitman, and Thoreau as principal contributions to the canon of American cultural history and criticism.

Like their British predecessors, the Young Americans launched a scathing critique of industrial capitalism. Their sense was that Europeans, though they suffered the same humiliation and ecological ruin caused by the industrial system, managed to hold onto some of the cultural treasures of pre-industrial life. This would prove to be more difficult for Americans. Americans would have to reinvent an entirely new community-based way of living centered on democratic self-reliance and unified by a shared culture. They saw community revitalization as kind of cultural renewal, and these new local cultures would be the seedbed of a broader political renewal. Significantly, Van Wyck Brooks expressed concern that such a movement would require Americans to create a new object of living; that is, an object of living that is not mortgaged to a consumer culture and is one that aspires to a more fulfilling life.

Brooks asserted that capitalism had reduced everything to market prices and self-interest. There were few if any chances to find self-fulfillment in a system that had little meaning beyond the self-interested pursuit of consumer goods and profits. Brooks writes, ". . . one cannot have the expression of personality so long as the end of society is an impersonal end like the accumulation of money." Such a system leaves Americans, according to Brooks, ". . . cold and dumb in spirit, incoherent and uncohesive as between man and man, without community in aim or purpose . . . a prodigious welter of unconscious life, swept by a groundswell of half–conscious emotions."[40]

Nearly a century ago, Randolph Bourne railed against a general condition of alienation created by the soulless marketplace and the swamps of corporate commercialism. All the standards of business success in a capitalist society—market penetration, profits, and the accumulation of money—could not be achieved, according to Bourne, ". . . without a certain betrayal of the soul." He also saw that people seeking to live with dignity were left with little choice but to be consumed by large corporate interests or ride out the rickety and uncertain vicissitudes of small business and wage labor. American life, according to Bourne, had become rudimentary and uninspired. Rephrasing Marx and Engels, Bourne proclaimed, "The world has nothing to lose but its chains—and its own soul to gain." With stunning prescience he asserted that:

> Those who came to find liberty achieve only license. They become the flotsam and jetsam of American life, the downward undertow of our civilization with its leering cheapness and falseness of taste and spiritual outlook, the absence of mind and sincere feeling which we see in our slovenly towns, our vapid moving pictures, our popular novels, and in the vacuous faces of the crowds on the city street.[41]

Bourne opposed modern capitalist society not because of a lack of basic rights or liberty, but because, ". . . it warps and stunts the potentialities of society and human nature."[42] Socialization within capitalism keeps people enslaved to the ever changing fads of consumerism, which for Mumford was replacing a fulfilling life with "the good life."[43]

Not long before the great stock market crash in in 1929, Waldo Frank wrote, "America as a whole is saturated with the vision and values of Wall Street." He attacked American consumer culture as self-obsessed, and herd that mentality of fashion posed as a hideous substitute for genuine solidarity among members of a community. Frank viewed social cohesion in 20th century American culture as the willing complicity of consumers with their own manipulation by big business.[44]

Though they were critical of capitalism, the Young Americans held an ambivalent view of socialism and did not look to national political movements as the sponsors of social reform. They sympathized with socialist ideology, but they could not see how a socialist movement would be anything more than another variation of nationalistic militarism and the concentration of industrial power. State socialism, in their view, was a mere changing of the guard.

They saw it as something that would only perpetuate the social and ecological destructiveness of industrial capitalism. Like Ruskin and Morris before them, the Young Americans did not seek a radical change in industrial relations, but rather to begin creating something wholly new and different. On this Mumford writes:

> The conclusion I drew for myself was that the situation demanded, not specific attacks on specific evils and points of danger, but a wholesale rethinking of the basis of modern life and thought for the purpose of eventually giving a new orientation to all our institutions.[45]

The Young Americans also did not see solutions in national politics. The progressives of their generation were still holding to the belief that they could create a new world through the politics of statecraft. Mumford in particular saw that large scale reform movements, uprisings, or sweeping pieces of national legislation demonstrate a fundamental ignorance of the organic relationship between the inner lives of individuals and their social setting. For Mumford understanding that relationship is centrally important for achieving true democracy.

The Young Americans envisioned democracy in a way that was quite different from the platforms of progressive political parties. For them, the goal of true democratic activism was not national legislation, but civic renewal through the revitalization of local community and culture. A local culture with a shared commitment to a productive and aesthetic life would become the medium through which individual personalities melded to form a broader political and social milieu. The Young Americans envisioned a more decentralized, community-centered model for a social economy as an alternative to industrial capitalism and the market system. For them, community is where people have a stable and secure residence and a sense of personal grounding or rootedness with a local flavor. It is a place where an individual has a whole life that, through a shared culture that is nestled within its natural environment, is unified with other members of the community. This is, of course, something that could never be legislated with referendums or statutes.

Their conception of a shared culture within a community was not sentimentality. Rather it was a meaningful and vital sharing of language, folklore, music, food, art, artisanal traditions, and a reverence for their natural surroundings. In this way, they were inspired by the Gothic revivalism of Ruskin, the Arts and Crafts Movement of Morris, and the organic-ecological visions of Geddes and Howard. Their vision of a Beloved Community was infused with an individual, moral, and even spiritual purpose. Their vision was that, by fostering the development of such a community, they would be setting the foundation for a radically new approach to democratic politics. As such, the Young Americans were meta-economists who held the innermost conviction that the ultimate goal of any political-economic system should be to support people as they strive to live full lives and to develop their own powers and capacities.

For the Young Americans this process of development begins with the essentials of life. What they see as essential is cultivating an understanding of the life-giving qualities of nature and refining the activities of the community: sport, theater, readings, festivals, religious rituals, musical performances, and the preparation and sharing of food—above all, nurturing the social relations among friends, family, and mates. They see these things as the primary object of living and the other stuff of politics and economics as secondary. In their view, we humans are an evolving species and are in a continual state of becoming. Our daily routines and duties at the workplace or office or farm are essential, but only insofar as they contribute to the ability of the people in the community to become something better—to evolve.

Their ideas on the connection between civic renewal and democratic institutions remained ambiguous, as the critics of the Young Americans were eager to point out. Like the pioneers of the transition movement, they left behind an unclear vision of institutional change. Personal development and public renewal need to be carried out in some fashion or another under the umbrella of a set of institutional rules, norms, and shared strategies that unite the community. The Young Americans' ambiguity on this was sufficient to marginalize their work as politically naïve. But their work did not wither on the vine.

By making personal development the central goal of civic regeneration and cultural renewal, the Young Americans highlighted the need for moral meaning and personal identity in the spheres of economics and politics. Progressives and Marxists of that time, though sympathetic with this view, tended to dismiss such things as irrelevant. The Young Americans also highlighted the notion of scale, which is to become centrally important in the discourse about the end of the Oil Age. Like the other meta-economists, the Young Americans lacked faith in national politics and rather saw that the emphasis of civic participation had to be on a smaller, localized or regionalized scale.

On this, Mumford writes, "Small groups: small classes: small communities: institutions framed to the human scale, are essential to purposive behavior in modern society."[46] In other words, small can be made beautiful. And to those who may be unfamiliar with it, reference to "small is beautiful" is homage to the work of E. F. Schumacher.

E. F. Schumacher, Economist and Philosopher

There are two dimensions to the work of E. F. Schumacher. One dimension is Schumacher, the British government economist; the other is Schumacher, the sage and philosopher. Taken as a whole, his work served as a guiding constellation for reinventing both our worldviews and our systems of production. The processes of transforming how we see and understand the world and the process of transforming our systems of production and technology are two core themes that run through the body of Schumacher's work. The bulk of his writings is contained in a collection of essays published in *Small is Beautiful: Economics as if People Mattered* (1973), *A Guide for the Perplexed* (1977), and *Good Work* (1979). The latter was published posthumously, as Schumacher died in 1977.

Like those of so many other of the meta-economists, Schumacher's ideas stood outside the margins of convention. Like them he was also critical of the standard dogma in the economics profession. So, naturally his work met hostility and skepticism from the pantheon of orthodox economics. Orthodox economics, though, while it attempts to dress itself up as an objective science, is normative ideology rooted in certain assumptions.

One of these assumptions is that anything that gets in the way of growth and accumulation is shameful. Another is the idea that growth and accumulation can continue for all time and people will never have to worry about scarcity of oil or other resources because, as the assumption goes, with higher energy prices in open markets, it becomes more cost-effective to develop and exploit energy alternatives previously considered too expensive. Dogmatic economists profess that with faith in the free market, new energy sources will always be forthcoming. As energy will always be renewed and available, economic growth can always continue.

Another assumption is the idea that self-interest and greed can be useful for economic development. This belief goes back to the 18th century and the work of Bernard De Mandeville and Adam Smith and continued on with the classical economists of the 19th century that were so reviled by Ruskin. The basic idea, as Ayn Rand celebrated in the 20th century, is that there is virtue in selfishness. Greed can compel affluence and create stark conditions of wealth alongside poverty. Economists see this as a necessary but temporary condition. It is believed necessary because we must allow the wealthy to amass their fortunes first as a process of capital formation. The capital then becomes the source for investment, growth, and then this leads to a virtuous condition of prosperity for all. With this assumption ethical concerns must be pushed aside and we must, in the words of 20th-century economist, John M. Keynes, ". . . pretend that foul is fair." In other words, we need not concern ourselves with ethics as ethics are, in fact, a hindrance. The belief is that the lure of personal gain will compel people, entrepreneurs specifically, to become industrious and productive. The entrepreneurial class then leads the charge for economic growth and the benefits of this growth will eventually accrue to us all. This sentiment was also expressed by former American president John F. Kennedy when he asserted that "a rising tide will lift all boats."

Schumacher makes a case for a more noble or enlightened approach to economics. He challenged both assumptions. Like Albert Bartlett, Schumacher made the common sense observation that the assumption of infinite growth in a world of finite resources is obviously impossible. As Chief Economic Advisor to Britain's National Coal Board for twenty years, he was in a unique position to see that the so-called rising tide would become troubled by oil scarcities, cartels, and high energy prices. Mainstream economists tend to ignore the problem of depleting resources because such an outcome does not mesh well with their cornerstone assumption that greed is good and ongoing growth is the salvation of us all.

Schumacher submitted that if people accept the notions that greed is good, foul is fair, and that vice leads to virtue, then our society will be molded accordingly into a culture of plunder. In such a culture there occurs a general collapse of intelligence, a loss of wisdom, and people become blinded by their own rapacity. As Ruskin warned a century before, people lose insight into their own nature and the reality of things around them. Schumacher saw that by accepting these assumptions, we have created a culture conducive not to ecological permanence, but to devastation and ruin. To plunder is to destroy and strip bare all that has value from other people and from our natural habitat. It is rape, spoilage, and ruin, yet rationalized by orthodox economics by the stroke of a pen: "foul is fair."

This rationalization easily settled in the popular imagination of people living in capitalistic cultures. Schumacher argued that in these cultures people, in large numbers, are likely to replace intelligence and serenity with greed and envy. He writes, "If whole societies become infected by these vices, they may indeed achieve astonishing things, but they become increasingly incapable of solving the most elementary problem of everyday existence." Nonetheless, this sanctification of greed still remains as one of the core tenets underlying orthodox economics. To most critics, however, this appears as nothing more than vulgar apologetics for business profits and accumulation. But for Schumacher this was symptomatic of a much deeper metaphysical crisis. It is here that Schumacher reads less as an economist and more as a philosopher.

For Schumacher, the dubious claims of orthodox economics represent an extension of a much deeper crisis, or defilement, of societies' innermost convictions—our metaphysical core. These convictions are the center from which the ideologies surrounding all scientific disciplines emanate like rays of the sun. For Schumacher, this core, or what he called the "eye of the heart,"[47] was traditionally grounded in the teachings found in all the great spiritual traditions, such as Christianity, Judaism, Islam, and Buddhism. In these traditions, this core is cultivated from not just from knowledge, but from deep insight and wisdom. For Schumacher, the "foul is fair" maxim is the antithesis of wisdom.

The eye of the heart, filled with this insight and wisdom, informs the mind and liberates the soul. It is that which gives the spiritually informed person the power to recognize and discern truth from fact. In all the great spiritual traditions, love, compassion, reverence for beauty and nature have been elevated above the world of facts. For Schumacher spiritually informed wisdom is what allows us to live with each other and with our environment and to develop ethical principles to guide our behavior and institutional development.

Schumacher was concerned that the ideas grounded in the great spiritual traditions have been gradually purged from this core and replaced with what he refers to as "materialistic scientism." In the scientific community, spiritual traditions have been largely dismissed as authoritarian megalomania or preachy dogma and have been replaced by the mechanistic constructs of the 19th century. This is particularly so in economics. For Schumacher, the result was that the metaphysical core became filled with cynicism and life-destroying and soul-destroying confusion. For him, this is a kind of spiritual emptiness and a belief system that lead people away from compassion toward cynical axioms of self-interest, survival of the fittest, and predatory competition. These axioms further justify violence, aggression, and a self-aggrandizing culture fraught with delusions of power and grandeur.

In his view, the meta-economic vision of materialistic scientism blinded the eye of the heart and left us adrift without insight and wisdom. He admitted that such a vision was appropriate for understanding the workings of the inorganic world and engineering, or what he called, "science for manipulation." But it is wholly inappropriate to humanity and its capacity for self-reflection and spirituality. By holding ourselves in a paradigm that consists of moving, mechanical parts we are mirroring ourselves as inanimate objects—we have lost our understanding of what it means to be human. By doing so, we undermine our ability for individual and cultural development.

As Schumacher saw it, we are destroying the land, climate, and ourselves not because our technology is flawed, but because the innermost convictions that we hold to be unassailable have been defiled. He is critical of the core concepts that have solidified into our metaphysical center and this center is what informs and guides all forms of inquiry. More specifically, for Schumacher, materialism has purged wisdom and higher-level truths and has allowed our cultures to be debased with the idolatry of wealth, fashion, self-aggrandizement, and technological gadgetry. From this core emanates a vicious ecological attitude toward our natural habitat which sees it as nothing more than a quarry for exploitation—a transgression against future generations.

Schumacher concludes, "The idea that a civilisation could sustain itself on the basis of such a transgression is an ethical, spiritual, and metaphysical monstrosity. It means conducting economic affairs of man as if people really did not matter at all."[48] Schumacher saw this as a crisis of how we perceive the world. If so, then the solution has to lie, at least in part, in changing our worldview. If through our own efforts we can change our habits of thought, we can break from our attachment to continual growth and self-interest. With a different mindset perhaps people can go about building new systems. Perhaps these first steps would be something like small-scale civic regeneration projects as envisioned by the Young Americans.

This process of changing how we see the world and how we act in the world is a core theme that runs through all of Schumacher's work. For Schumacher, the reason that the core materialistic beliefs can be destructive is that they lead to the development of life-destroying technology. Though so much of what Schumacher emphasizes revolves around core belief systems, he contends that the starting place for survival is with simultaneously changing material technology and our core belief systems. Material technology and industry are locked into a dynamic, self-reinforcing interplay with our belief systems. That is, we cannot change our habits of science and engineering without also changing our worldview.

A shift toward a more spiritually informed metaphysical core would be more compatible with what he refers to as "technology with a human face." For Schumacher, the goal of technological development ought to be the lightening of our burden and enhancing the joy of work. Technology has not done this. It has lightened the burden of work in certain places by making it heavier in others. Tremendous systems of mass production remove the intrinsic joy that good work can bring and replace it with meaningless, joyless, assembly line toil, and deskilled labor in sweatshops.

Schumacher's ideas about technology with a human face originate from a sense of compassion for the poor and unemployed people in the so-called third world. He observed that their development models were essentially copying the life-destroying industrial systems that have long since been developed in the West. As an alternative development model, Schumacher offers "intermediate" and "appropriate" technology. His proposal for this technology is rooted in four basic propositions.

First is to decentralize industrial systems and move work to where the people already live. Rather than the giant megalopolises, he envisions a decentralized system or network of "two million villages," a network of small semi-autonomous mini-systems. Second is to develop inexpensive technology that will not require large amounts of capital formation for investments, which is a resource that is very scarce in the developing world. Third is to design relatively simple production methods that are easily implemented and maintained. Fourth is to implement production systems based on local resources and directed toward local use and consumption.[49]

Technology with a human face for Schumacher is a development model centered on bringing simpler, less costly, less resource consuming, and easy to maintain tools and equipment into the towns and villages where people already live and work, rather than giant factory systems in highly concentrated urban areas. The systems are flexible, adaptable, easy to maintain, decentralized, small scale, more manageable, and democratically accountable. The goal here was originally directed to absorb poor, unemployed people into meaningful and productive work and to alleviate poverty. Schumacher wanted to see these ideas implemented and thus founded the Intermediate Technology Development Group in the 1960s.

Though this was a model originally intended for developing countries, it became appropriate for creating more sustainable alternatives in the so-called developed nations as well. Over the decades since Schumacher was living and working his ideas took root. Alternative development models based on his principles were implemented in both developing and developed worlds. His ideas provided inspiration for the many movements around the world that are directed to developing community-based economics and voluntary simplicity.

Toward the end of his life, he left us with a vision of a future that is visible to us now. Schumacher the economist tells us of a future economic system that exists as a patchwork of semi-autonomous smaller-scale organizations both public and private. With a different mindset people could re-create economic systems governed by rules and codes that lead to very different outcomes—ones that are governed democratically through worker- and community-based councils. Like Ruskin and Morris, he had a vision of production systems that employ the skills of people who find actual joy in their work and at the same time innovate technology that is appropriate to lighten the burden of the workers yet does not take away their crafts. Appropriate technology means improving the mastery of simpler methods of production and downgrading those that are excessive and overly complicated. He warns that what precisely is appropriate technology is difficult to pin down. There is no one-size-fits-all model. The technology is always developed within local institutions and local cultures. Most important is that it must be developed on a grass roots basis situated in the local culture and environmental habitat. Food production is by and for the local population. Smaller, simpler production systems naturally are more flexible and can be adapted to changing environments. He also envisioned a future social system of production that is grounded also in sound ecological principles.

Conclusion

The question is not whether or not we will return to smaller, simpler localized systems of production. A future without fossil fuels will make sure of that. The question is whether or not we can make those systems beautiful. With these meta-economists as our inspiration, we can ascribe meaning to the word "beautiful." Our great transformation of renewing our local economies and communities is the process of developing smaller scale production systems that are community-based, values-based, and more democratically self-reliant; that are held together by a vibrant local culture; that integrate aesthetic creativity with work skills and appropriate technology; that connect self-fulfillment with healthy lifestyles and a revitalization of public life through local citizenship; that defend human dignity and imagination; that provide economic stability and fairness by structuring ownership in the community and protecting people's livelihoods that are situated within the carrying capacity of their natural habitats.

Of course you can imagine the cynics are ready with curled lips to make their snarling comments about our "ideals, ideals!" But this did not stop these meta-economists from creative thinking. It is a far easier task to play the role of the cynical critic than the visionary, for the cynic does not have to use his imagination. It easier to attack someone for having ideals that differ from the conventional pieties for the obvious reason that the attacker has the established way of thinking and doing things—no matter how destructive or pathological—on his side. The visionary has to be creative and imaginative, whereas the cynic merely needs to point out the obvious fact that a true alternative faces the difficulty arising from being different from the *status quo*.

When Ebenezer Howard's Garden City Association launched its first garden city project in Letchworth, the planners organized exhibitions that would showcase their work. They hoped that if people saw what they had accomplished, the movement would gain momentum. The exhibitions attracted tens of thousands of visitors, including the derisive critics who mocked and scorned. One columnist sneered at those involved as a bunch of "idealistic health freaks." To the critics, Howard responded with some aplomb:

> What Is may hinder What Might Be for a while, but cannot stay the tide of progress. These crowded cities have done their work; they were the best which a society largely based on selfishness and rapacity could construct, but they are in the nature of things entirely unadapted for a society in which the social side of our nature is demanding a larger share of recognition . . . The reader is, therefore, earnestly asked not to take it for granted that the large cities in which he may perhaps take a pardonable pride are necessarily, in their present form, any more permanent than the stage-coach system which was the subject of so much admiration just at the very moment when it was about to be supplanted by the railways.[50]

The meta-economists have given us rich food for thought. But vision and ideals have to be put into practice if we are to get out from under the ecological and economic messes we are in. Putting them into practice means elevating our core beliefs about what is good, beautiful, fair, or wholesome and making sure that our economic system allows us to live according to these beliefs. This is a tall order because it would require that we build a new, shadow economy out of entirely new institutions with new rules, norms, and shared strategies for social behavior.

The process of crafting new institutions is largely missing from the meta-economists' suggestions in the same way that it is missing from the transitioneers' suggestions. Filling this gap—the nuts and bolts of crafting new institutions that can develop into a new social economy or commonwealth—is what I'll examine in detail in the next chapter.

CHAPTER FIVE

THE NEW MONASTICS

The meta-economists gave us more than just the vision that our work can foster personal development, give us a sense of dignity, and be a source of community renewal. They provided inspiration for future iconoclasts to stay resolute with their visions for an alternative future despite the resistance they face from established conventions. The need for such vision and inspiration is perhaps more crucial today than any time in modern history as energy descent signifies the descent of business as usual based on oil.

Since the 1970s and perhaps even earlier we have been warned by Schumacher and others about the dangerous folly of allowing our economic system to be so centered on this single, one-time, finite resource. Though these warnings have been clear and compelling, they have been largely ignored and our efforts to kick the oil habit remain lethargic. We cannot change the immutable laws of physics and chemistry that made oil a finite resource, but we can change our economic institutions. The longer we ignore the reality that real institutional change must be achieved, the harder it will be to do and the more perilous our world will become. It is unwise of us to ignore the dangers of our oil dependency, just as it is unwise for us to ignore the lessons from our own history—particularly our mistakes.

In the American experience, perhaps the last great historical transformation of this magnitude was the end of slavery. Yet, neither before, during, nor after the Civil War did those in positions of power—North or South—give serious consideration to creating new institutions that would replace that "peculiar institution" so intertwined with racial hatred. During the period of Reconstruction, the *status quo ante* between whites, blacks, and the land remained largely unchanged and blacks were blocked from pursuing true economic independence as equal citizens. Sharecropping and plantation wage labor that developed in the aftermath were little more than slavery in disguise. The habitual tradition of white supremacy lingered and was expressed in the oppressive Jim Crow laws that followed Reconstruction. Jim Crow laws institutionalized the segregation of whites and blacks, with blacks systematically shut out of the structure of economic, educational, and social opportunity, and these remained on the books until the Civil Rights Act of 1964 rendered them unlawful. Had there been better forethought, a true sense of justice based on human rights, and a vision for institutional change among those who were remaking history during the period of Reconstruction, American society would have evolved in a very different way. And had there been a genuine commitment to economically enfranchising the former slave population an entire century of troubled race relations could have been avoided.

The institutional exception was the black church. During the years of slavery, the black church and congregation were virtually invisible, ostensibly to prevent white paranoia about black gatherings into mobs. After emancipation, black churches moved into the light of day to become the central institution of black communities. On this, historian Eric Foner writes,

> Churches housed schools, social events, and political gatherings. In rural areas, church picnics, festivals, and excursions often provided the only opportunity for fellowship and recreation. The church served as an 'Ecclesiastical Court House,' promoting moral values, adjudicating family disputes, and disciplining individuals for adultery and other illicit behavior.[51]

In other words, the black church was the first social institution fully controlled by and for black people in America.

As we approach the end of the Oil Age, we can use our capacities of hindsight and foresight to avoid repeating our past mistakes—although we seem poised to do just that as we cling to our conventional ways of doing things. If we want to avoid another troubled century, we have to create new social structures by and for our communities such that they step outside the circle of business as usual. Without forethought and the hard work of crafting institutions that are cut from new cloth, the idea of "change" becomes a symbolic window dressing without meaning or substance. We do not have to create ecclesiastic social structures necessarily, though like the Southern blacks, we can preserve what is to be cherished in this otherwise crumbling system and its culture. In other words, we can become what cultural historian Morris Berman calls "The New Monastics."

The New Monastics and Crafting New Institutions

In *The Twilight of American Culture* (2000), cultural historian Morris Berman tells of the Franciscan monks and others in the 4th century who "took it upon themselves to preserve the treasures of the Greco-Roman civilization as the lights of their own culture were rapidly fading."[52] Berman draws a parallel between these monastics of early middle ages to what he calls the New Monastics. These are those in our contemporary society

> who resist the spin and hype of the global world order; he or she knows the difference between reality and theme parks, integrity and commercial promotion . . . dedicated not to slogans or the fashionable patois of postmodernism, but to Enlightenment values that lie at the heart of our civilization: the disinterested pursuit of the truth, the cultivation of art, the commitment to critical thinking, inter alia.[53]

The New Monastics are aptly summarized on the cover of Berman's book by Ferenc Morton Szasz, "[People] who are willing to reject the corporate consumer culture of our time and instead work to preserve the historical treasures of our civilizations."

For our purposes here, we can think of the New Monastics as those among us who will craft new institutions patterned after businesses, governments, nonprofits, and other social structures that are treasures worth preserving in our current system that we will leave behind. These are the organizations governed by rules that are unconventional in the sense that they place truly sustainable practices, community empowerment, aesthetics, and social justice over profits and growth. And although no single one of these organizations necessarily stands as a perfect model itself, taken together they constitute the raw material used to build new institutions for a new commonwealth—a new commonwealth for the end of the Oil Age.

In this way we can see those of us who are committed to working for a better future as New Monastics who have embarked on a journey to craft new institutions. The first phase in this journey is to establish a new ideological foundation upon which our new institutions will be created. Just as the original architects of permaculture and the transitioneers set out to re-localize economies based on certain core principles, our New Monastics need to have a set of principles that serve as the seed bed from which new institutional forms will grow and support each other. Here again we can follow in the footsteps of Schumacher and the other meta-economists who sought to develop wisdom, aesthetics, and a wholesome ideological foundation from which to build a broader social economy. The meta-economists have created for us an ideological foundation that comes from the heart and is based on that which is intrinsically health-giving, beautiful, just, ecologically permanent, and that which we will not allow to be systematically compromised. Foundations are set in concrete. Upon this concrete foundation we can build a new commonwealth.

In my earlier work I refer to this concrete ideological foundation as "mindful economics." The goal of a mindful economy is to create a new commonwealth based on a system of economic production that is directed toward serving the common good. In the Schumacherian sense, this commonwealth would be structured to function for what is stable, equitable, and ecologically permanent. The first step for the New Monastics, therefore, would be to establish a set of normative principles that will constitute the basis from which we develop our values-based institutions.

One starting place could be with the ethical principles of permaculture: (1) Meeting the needs of people in ways that are wholesome and genuine. (2) Meeting these needs in a way that is fair, stable, equitable, and genuinely democratic yet allowing for emphasis for those whose voices are squelched under the clamor of the democratic majority. (3) Meeting these needs such that we do not systematically damage our natural environment, which we interpret to mean not systematically damaging the carrying capacity of our planet for all variations of life.

And we can expand on these principles of permaculture: (4) Working to meet the needs of people while transforming households from acquisitive and obsessive consumers to integral institutions of an ecologically and economically stable community. (5) Restoring financial institutions to their original and core function of providing financial services for local development and evolving away from the expectation that financial markets are always going to pay off with exponential returns. (6) Making sure that markets and market exchanges are not allowed to be swallowed up in a vortex of greed and speculation, but rather are centers of vibrant community cultures. (7) Transforming local governments into truly democratically accountable institutions that are not captured by moneyed interests and from this basis building a populist platform to push upward and making state and federal government agencies similarly accountable. (8) Making businesses into institutions that meet the needs of people and communities as their primary responsibility, not the maximization of shareholder equity.

From this list of principles, which is obviously expandable or contractible, we can begin the process of crafting institutions. Institutional development is based on a process of what political economist and Nobel Laureate Elinor Ostrom refers to as "rule structured situations."[54] That is, the rules we live by are specifically structured to create a social environment that will lead toward some purpose. Our purpose is to create a re-localized, ecologically permanent, stable, aesthetic, and socially just commonwealth that will allow us to live mindfully in the absence of cheap oil.

Clearly this will be no easy task, but Ostrom gives us some guidelines for creating the appropriate institutions for managing a commonwealth. One is that the commonwealth itself needs to have clearly defined boundaries. This is not just a geographical delimitation, but also an identification of the community that is served by the economic activity taking place within it. Scale is important here. Keeping the boundaries close is helpful in maintaining close feedback and communication among producers and consumers, yet at the same time the boundaries must be broad enough to achieve some measure of self-sufficiency. In some cases the boundaries could delineate a specific neighborhood within a larger urban setting, or it could encompass an entire state depending on population concentrations and the availability of resources.

These resources must be appropriated and used in ways that are rule-structured. There are different dimensions to the concept of rules depending on how we want to frame our actions and choices. Living, working or playing within the confines of rule-structured situations is almost entirely what we all do on a daily basis. Some of the rules are formal with "or else" consequences and some are informal shared strategies. To give a simple example, the parents within a basic family structure, a core social institution, may establish a household rule that says, "Everyone in this family will brush his or her teeth before going to bed." Or a business may post a sign that says, "If you jam the copier, you are responsible for clearing the paper jam" or "Every item in this store is subject to a 6 percent state sales tax." On a neighborhood street with a school nearby there will no doubt be a sign indicating that the speed limit is 25 mph.

The underlying implication of these rules is that there will be consequences for those who do not follow them, although those consequences do not necessarily need to be stated explicitly. And of course there is the rule of law.

Ostrom calls our attention to what she refers to as the grammar and syntax of institutional rules. That is, the rules we live by are expressed as institutional statements. In the US Constitution, for example, there is this statement that abolishes slavery:

> Section One. Neither slavery nor involuntary servitude, except as punishment for crime whereof the party shall have been duly convicted, shall exist within the United States, or any place subject to their jurisdiction. Section Two. Congress shall have the power to enforce this article by appropriate legislation.

Institutional rules can be less formal and take the form of shared strategies. Individuals who share a vision, a set of values or goals might adopt rules or guidelines without codifying them formally. A shared strategy that lacks specific "or else" sanctions is more likely to work as an informal institution and can be just as effective as formal institutions if it garners enough support, has enough legitimacy in the eyes of community members, and can exact compliance. That is, advising cooperation or compliance to achieve a goal is equivalent to informal institution rules rooted in a community that share a common set of values. If a sign posted at the entrance to a hiking trail in a park says "The rule of the trail is to leave it as you found it," it does not have specific "or else . . ." sanctions, but stands more as advice to cooperate so as to keep the park in good ecological condition. Shared strategies do not necessarily need institutional statements. There was no statement advising people to people carry cloth bags to their grocery store rather than using more ecologically damaging paper or plastic bags. The social practice of carrying cloth bags emerged from a shared strategy rooted in a shared ecological vision and set of values.

Ostrom also emphasizes the need for collectivity in setting community rules. To be just and democratic the institutional rules of our commonwealth must be established collectively through participation by the members. Whether they are formal laws or shared strategies, the rules will not serve the common good if they are arbitrarily imposed on the community by a small elite or dominant group.

Most of us find that these rules are relatively easy to follow. We have been following so many of them that they simply become part of our vast repertoire of habitualized behaviors or cultural norms. As they become more deeply entrenched community members no longer need to be reminded of the rules as they just become part of daily life. Institutional rules are as ubiquitous as the air we breathe. They script how we interact with our families at home, how we drive in our cars as we commute back and forth to work, how we interact with our coworkers or boss, the way we recreate or participate in social activities, how we play sports or music, and how we shop. So when we talk about creating new institutions, we are not talking about imposing new rule-structured situations on people, those structures already exist.

The economic institutions that comprise this commonwealth do not exist in isolation. They are layered together into a complex web of interconnected ions that form higher-level systems and culture. Part of the reason we adopt a broader systems view of our economy is that we have to know we go about creating new institutions and new re-localized common- , we need to be continuously aware and mindful of the fact that no how subtle or imperceptible the changes we make seem to be, we are g our whole society. As we craft new institutions we are changing our e.

s," as Ostrom warns us, "understanding institutions is a serious r." And, again, underlying all of these institutional rules is a set of The crux of all of this is that if the rules, norms, or shared strategies live by are clashing with our core values, then it will be up to us to the rules. The paramount task of the New Monastics, therefore, is to ew social structures governed by a new set of rules and shared strate- t are consistent with the core ethical principles underlying the designs new commonwealth.

eat transformation will not be an attempt to go out and completely l our economic system or culture. That is not possible. It is rather utionary process of change in which we use the most important tools we have available, our conscious minds and will, to transform our habits of thought and action in the direction of something new. Perhaps the best way to begin our great transformation is by opening space for thinking creatively about the kinds of institutions we will construct and how those institutions will weave together into our re-localized, community-based commonwealth.

For this effort, there are models that serve as templates such that we don't have to reinvent the wheel.

The case studies in the subchapters that follow combine to form a kind of dream team of living models. Each contains treasures that can be collected from our current dying and dysfunctional system to be carried forward by the New Monastics as the seeds of a future commonwealth. And each is systematically connected to every other. These studies are of actual business models, government programs, and other initiatives that have promise to lead our economic activity in a different and better direction. What makes them stand out from business as usual is that they are organized around a different set of rules and shared strategies conceived from the ideas that reflect the list of ethical principles necessary for social, ecological, and even spiritual well-being.

As I mentioned above, no single enterprise or organization that I cover here is itself perfect, but together they sum to an image of a very different system of production. Recall that Patrick Geddes's vision for civic regeneration was to create a new eutopia, a civic society that has the potential to exist when all the actual, real world conditions for its existence are met. Connecting the treasures from these living models is a way to evolve our economic system toward something that would realize Geddes's dream. It is impossible to say exactly what this commonwealth will look like until it emerges, nor will we have a new "ism" to define or categorize it. But we should not be daunted by this uncertainty and have faith in our true values, cultivate the eye of the heart, and allow the eutopian commonwealth to evolve into being on its own accord.

The B Lab Connections

Our first treasure is a business model based on the idea that corporations can be crafted to be agents of change. Much of what defines a corporation's activities depends on how it is originally incorporated. The conventional view of a corporation is that it is a business institution incorporated specifically as a vehicle for profit maximization. The founders of B Lab, however, have a different view. Through their work the people at B Lab hope to unlock investment capital to be used, unlike conventional corporate institutions, to have a material, beneficial impact on society and the environment.

B Lab is a nonprofit organization started by Jay Coen Gilbert and Andrew Kassoy of the progressive leadership think tank The Aspen Institute. Their aspiration is to change the economic system from the inside out by certifying a new breed of corporations, B Corporations®. B Corporations are business models based on the principle that a corporate entity can be a blank slate and its charter can be written to be a force for social and environmental change; that is, a business that is also a force for social equity and environmental sustainability. The B stands for benefit—social and environmental benefit—and as such fits comfortably into the commonwealth. Although it is a relatively recent development, it has some potential to become a force for real long term change.

The primary initiative of the B Lab is to help existing companies become B Corporations through a rigorous certification process. Becoming B certified is comparable to getting products certified as organic or fair trade; however in this case it is not the products or services that get certified, but the company that produces them. The first step in the certification process is for the company to enroll in B Lab's B Impact Assessment. The assessment is a questionnaire that has evolved from other more traditional models for evaluating socially responsible investing such as GRI Conscious Capitalism and others. Most traditional socially responsible investment assessments are based on a single issue such as their environmental record or profile in fair trade. B Lab expanded from these models to assess a broader scope of issues including social equity, community responsibility, and environmental sustainability. The nature and structure of the questionnaire varies depending on the kind of business with a more rigorous rubric for manufacturing companies as they have greater potential for environmental impact than those in the service sector. Overall the assessment is a survey that covers several categories: corporate governance, workers, community, environment, type of business model, and certain disclosures.

On governance the assessment asks questions about whether or not the company has integrated its commitment to social responsibility and environmental stewardship in its broader corporate mission. The assessment probes to find out if employees are trained in this corporate mission, if there are appropriate accounting and reporting methods, and if it has clearly-stated policies to protect whistleblowers. In the section for workers, the assessment checks for the ratio of part-time to full-time employees, wage levels, bonus and profit sharing plans, worker ownership, health insurance, and the quality of the physical work environment. It asks questions about employment standards for females, minorities, disabled, or other previously excluded categories of employees.

The assessment also contains metrics for gauging the company's performance on its support of its local community. For example, it makes inquiries about whether the suppliers or owners are local, which for B Lab means within a 200-mile radius of the companiy's headquarters or main production facilities. It also checks the social and environmental track record of all the outside firms integrated into its supply chain including suppliers as well as end users. It also asks what kind of banks it works with, what percentage of jobs it has created for people located in low income communities, and what community engagement activities or charitable giving programs it has.

B Lab assesses the company's environmental impact in depth. It examines the business's practices relating to renewable energy use, greenhouse gas emissions, water use, hazardous or toxic waste production, recycling, use of biodegradable or environmentally preferred materials, environmental reviews or audits, LEED certifications (Leadership in Energy and Environmental Design) for its buildings, energy conservation efforts, and transportation and distribution efficiency.

The survey also checks for specific types of business models. Is the business a social enterprise created specifically for the purpose of addressing economic inequality, health, culture or some other social or environmental purpose? Was it created to rebuild part of the local community? How well integrated is the business within its community through local ownership? Is it a producer or artisanal cooperative? Does it exist for charity, to alleviate poverty, or to protect land and wildlife? The survey also looks for information on how well the company has stayed the course with its overall mission.

As a final piece, the assessment asks that the company make certain disclosures. To be certified a company has to disclose whether or not it is engaged in the production or trade of products or services that are related to something that may be potentially harmful. This includes producing alcohol, logging or logging equipment, weapons, genetically modified organisms, mining, nuclear power or fossil energy, tobacco, pornography, products related to endangered species, animal testing, child or prison labor, workers' rights, workplace accident records, worker relocation, or records on fines or penalties related to the company's activities.

From the company's performance in these categories a composite or overall score is derived with a total possible of 200 points. If the company scores 80 out of the 200, or 40 percent, it meets the B Lab's minimum standard for certification. At first glance, 40 percent may not exactly resonate in the imagination as flying colors for an impact assessment that vets and verifies operational standards for a business. But with a closer look it becomes clear that the B Lab Council has created a scoring rubric that is much broader than a single-issue test. As they make the criteria for certification more comprehensive, the company profile for social and environmental benefit gets stretched thinner. So the 40 percent standard was derived to be rigorous as it covers a wide range of social, environmental, and community concerns, but at the same time one that is possible to achieve. Once the company completes the assessment, B Lab follows up with a phone interview for clarification and random inquiries for substantiation. The purpose here is to check for accuracy and veracity in the questionnaire responses.

For this assessment service, B Lab charges on a sliding scale ranging from $500 to $25,000 depending on the size of the company. Moreover, to maintain their certification they have to repeat the assessment every two years. According to the people at B Lab, it is in the company's business interest to maintain these rigorous procedures. The assumption is that B certification stands as a prestigious social and environmental seal of approval by third party standards. And once a company becomes certified it can be seen more favorably by socially and environmentally conscious investors. This can also elevate the value of the company if it is considering selling to another business, and it can help the business raise capital.

Once the company completes the assessment process and the assessment is verified, the next step in the certification process is to have the company incorporate its benefit mission into its legal documentation. It is here that we get to the heart of institutional change. Insofar as economic institutions are social structures that script economic activity with rules and shared strategies, B Lab puts the rules and strategies of social and environmental change into the DNA of corporate institutions. Depending on how and where the company is organized, this would involve amending its Articles of Incorporation or Partnership Agreement to include the social and environmental responsibilities of the company that will have standing along with the conventional fiduciary concerns.

Meddling with fiduciary concerns can raise legal issues. Dealing with legal aspects of this force for change is another one of B Lab's initiatives. As corporate charters are still controlled by state governments the process of creating the B Corporation as a corporate entity requires some dealing with the state, though the extent of it will vary from state to state. The key issue is that a B

Corporation can be certified according to its own rubric, but whether its new certified mission will hold up against shareholder activism in how courts generally deal with corporate law is another institutional issue.

In a world dominated by capitalist institutions sovereignty over corporate governance in the view of most courts rests primarily with ownership. In the case of a typical corporation, it would be the common shareholders who are primarily granted the right to expect the corporation to maximize shareholder value. So if a B Corporation places social and environmental responsibility into the Articles of Incorporation or Partnership Agreement, and that responsibility compromises shareholder value, shareholders could have grounds to sue for damages wrought by disenfranchising shareholders and for violating the company's fiduciary responsibilities. There are, however, "constituency laws" in over thirty states that allow corporate directors to consider the interests of others such as employees, suppliers, members of the local community, and even customers. Once a benefit corporation is established through the legislative process, it is defined to pursue these broader responsibilities and those responsibilities will have legal standing.

B Lab provides both new and established companies a sense of direction that they need to follow in order to truly become a force for positive change. It also has a Global Impact Investment Rating System (GIIRS, pronounced as "gears" by B Lab people). GIIRS is an impact investing tool for meeting social and environmental criteria that helps investors determine the effectiveness of a company or fund in meeting these criteria. B Lab encourages businesses to not simply adhere to social and environmental guidelines, but to be proactive as agents of change. The mission of social and environmental benefit is structured into the specific institutional language—institutional statements—that govern the enterprise; what the people at B Lab call the enterprise's "legal DNA." This constitutes a process of advancing from Elinor Ostrom's concept of creating new "rule structured" situations that are central to institutional change.

As a force for institutional change, B Lab is relatively new. B Lab has been certifying companies only since 2007, which in the span of peak oil or climate change time is but a brief moment. But in that time, B Lab has certified over 500 companies representing over $2.5 billion in business revenues spanning sixty industries. The certification process is not a perfect guarantee that such companies are going to bring about the kind of changes we need to survive in a world without oil. However, unlike conventional corporations they are living examples of institutions that have positive change as part of their corporate responsibility and transcend profit for profit's sake or growth for growth's sake. As such they are treasures that the New Monastics can carry forward and help generations of forward-thinking entrepreneurs and investors build wholly new institutions that are created for positive change.

With each B Corporation certification, the company gets to craft its own statement about "the change we seek®." on the B Lab's final report. Here are some examples:

Build the best product, cause no unnecessary harm, use business to inspire and implement solutions to the environmental crisis." (Patagonia)

Namasté Solar's mission is to propagate the responsible use of solar energy, pioneer conscientious business practices, and create holistic wealth for the community. We infuse this mission in everything we do." (Namasté Solar)

... to be a leader in the constant efforts toward a just, sustainable world through excellence in their daily work in a vibrant, socially-responsible enterprise that is dedicated to creating & promoting meaningful products, built to high standards of craftsmanship, utility & sustainability." (AlterECO)

The long range implications of their work could be profound. Such a comprehensive rubric for business performance could evolve beyond corporate governance and rule of law to be established as cultural norms and concretely habituated in business practices. If such standards for social and environmental benefit become a cultural norm in the business world, perhaps the destructive habits of our past will be broken. Obviously much more needs to be said about breaking out of conventional modes of thinking and this is particularly true of how we think about money.

Banking and the North Dakota and Santa Fe Connections

It sort of goes without saying that monetary institutions are centrally important aspects of any economic system. Money is more than just a medium of exchange. It is an institution that connects people to one another. In that capacity it can either form the bonds that can define a community or it can tear the community apart. Over the last few years the pathological condition of our banking and financial system has come into full view as if someone had taken the lid off of Pandora's box and allowed every possible dismal thing you could imagine to escape into the financial sector. Moreover, the conditions for further instability are getting worse. What has emerged from this mess is one of the most concentrated industries in the world in which a handful of giant banks dominate the banking industry. To consolidate their dominant position, these banks have pulled the US Treasury and the Federal Reserve into their inner circle that remains enshrouded in a toxic cloud of ethical opprobrium.

People everywhere are looking for alternatives to this undemocratic and corrupt structure. Heading up the popular backlash against the too-big-to-fail banks was Arianna Huffington's Move Your Money campaign. Huffington and others at the end of 2009 made New Year resolutions to yank their money out of the leviathan banks and put it someplace else; presumably someplace that is smaller and more ethical. Many followed the call and deposits rallied at local credit unions, community banks, and savings banks. The populist boycott was an attempt to kick big, arrogant banks in the shins by making them lose customer deposits. The campaign managed to move about $4.5 billion in deposits by the fall of 2011, but compared to the trillions controlled by the big banks the campaign made not the slightest dent in the bankers' shins.

These short-lived boycotts are more expressions of frustration and powerlessness than effective movements toward real change. Like buy nothing days, or buy no gas days, or even the going green and local consumer trends, these movements suffer from a kind of delusion of grandeur regarding the sovereignty of the consumer. It would be nice if all we had to do was make different and better choices with our money, but it is just not that simple. Credit unions and community banks were originally created as social ventures or to meet the needs of their local communities, but not to take over the main thrust of America's banking industry. Our endeavor to build a new commonwealth comprised of distinct corporate and financial institutions is not a matter of simply making choices, though as choices become increasingly limited with industry concentration, it also is a matter of scale.

Credit unions are very small scale not-for-profit institutions. Unlike nonprofits which rely on donations and grants to cover operations, credit unions' cash flow comes from charging interest and finance fees though they do accept donations. Their mission is not to amass or accumulate fortunes, and their 501(c)(14) tax-exempt status allows them to generate only a small amount of surplus, nor can they sell shares to the general public. These restrictions prevent them from generating capital assets like commercial banks, so they are destined to remain relatively small. There are about 7,300 federally insured credit unions with about 90 million members and a little over $900 billion in assets. Community banks are small locally owned and operated banks that are chartered to serve families, farmers, and small businesses in their local communities. There are close to 8,000 community banks in the US with a combined total of about $1 trillion in assets. So again, to give a sense of scale, the total assets of 15,000 or so of these tiny depository institutions combined are still less than those of the one largest leviathan: J. P. Morgan Chase. If Americans were to move all of their money out of the four largest banks, they would have to build an additional 40,000 small banks just to absorb the deposits.

The point here is that if we are serious about creating a more democratically accountable, re-localized economy, it is going to require much more work than merely going online to shop for a new customer friendly community bank or credit union. Democratizing our banking industry and scaling it down to the community level will require a vast and strategic effort. To that end, we have much work ahead of us. And as we start this work of crafting new financial institutions to meet the needs of our commonwealth, we'll need good models; that is, some treasures for the New Monastics. I found such a treasure in Santa Fe, New Mexico: the Permaculture Credit Union.

The Permaculture Credit Union (PCU) is a small financial cooperative with slightly more than $5 million in assets, which are mostly loans to its members. Like other credit unions, PCU is a member owned, not-for-profit bank and is federally insured by the National Credit Union Administration (NCUA). The NCUA itself is a federal government agency that regulates and charters federal credit unions and, like the FDIC for commercial banks, it insures deposits up to $250,000 per account in all federal credit unions and a majority of those with state charters. What makes the PCU different is their institutional statement, "The PCU is a credit union dedicated to the Ethics of Permaculture: Care of the Earth, Care of People, Reinvestment of surplus to benefit the Earth and its inhabitants." The bank was founded in 2001 after two years of evolving from an idea to getting a charter. Its members have placed about $4.8 million in equity into the bank through membership share certificates and member savings. Nearly all of that money is lent back to the members for loans that are made to finance projects that are consistent with the Ethics of Permaculture such as solar energy, water catchment, retrofitting older cars to improve mileage or for alternative fuels. The vision held by the founders of PCU is not just ecological. They see the importance of recycling the dollars within the community to sustain the vibrancy of the local economy.

Given the restrictions placed on them for raising capital, PCU has proven resourceful. They have been able to attract capital from community donations channeled through an affiliated nonprofit, the Permaculture Guild. They are also connected to other financial institutions such as RSF Social Finance, US Central, and what is known in the credit union industry as the "26 Corporates." RSF is an equity fund that was established in 1936. The fund was established to address problems related to food production, education and the arts, and ecological stewardship. It provides loans and grants to both for-profit and nonprofit social enterprises that are social ventures attempting to foster progressive change. RSF has been a mentoring institution to PCU and helped them aggregate over $300,000 in badly needed capital.

US Central is a wholesale financial institution established under the auspices of the NCUA to provide funding and investment services to retail credit unions. It is the largest credit union in the United States with about $45 billion in assets and, as the Federal Reserve is the bankers' bank, it is the credit unions' credit union. It works through a consortium of twenty-six fairly large scale corporate credit unions. The 26 Corporates handle the larger, complicated financial transactions that are beyond the scope of the small, community-based credit unions. In this way, the credit union is structured into a stable pyramid with US Central at the top functioning as a kind of central bank, the twenty-six Corporates, and 8,000 retail credit unions serving the needs of their communities. Part of what makes this structure stable and functional is that it is entirely anchored to its original mandate of providing financial services for economic and community development. This stands in stark contrast to the commercial and investment banking industries that are anchored to the pathology of speculation and corrupted by the moral hazard that comes from the knowledge that they can be as reckless as they choose because the Fed and the Treasury are on standby ready to bail them out.

The stability, resourcefulness, and genuine commitment to community wellbeing of this structure can go a long way toward creating a healthy, functioning re-localized commonwealth. This can be further augmented by developing within each commonwealth a public option—a public bank. Like a public school or a public utility, a public bank is state-owned and operated.

The only state-owned bank of any scale currently in operation is the Bank of North Dakota (BND) located in Bismarck, North Dakota. Despite the paranoid and shrill condemnations of fundamentalist libertarians and right wing radio pundits, the BND is a remarkably stable and crucial part of North Dakota's economy. Since the latest turmoil in the banking sector began in 2008, BND has been heralded as a viable alternative and several states are seriously considering following North Dakota's lead.

Established in 1919 by North Dakota's state legislature, the BND is a correspondent bank, sometimes called a partnership bank. It was originally created as an alternative to the predatory and poorly regulated private bank monopolies that dominated the industry at the time—a time that we have returned to nearly a century later. It was chartered to help improve the economy of the state by assisting the development of local financial institutions. Like US Central in the credit union business, the BND serves as the bankers' bank for community banks located within the state. It is a for-profit institution, though aside from small student loan and business development lending programs, the BND does not make loans directly to borrowers. Rather as a correspondent bank it is chartered to provide banking services for other financial institutions.

The source funds for the BND come primarily from the state treasury. By law, state government agencies in North Dakota are required to deposit their funds in the BND, though local municipal agencies are exempt from this requirement. Outside of North Dakota virtually all state and federal government agencies maintain their treasury funds in special accounts with private commercial banks. The private sector is not required to keep their funds within the state and can be used outside the state either for legitimate purposes or for predatory or speculative ventures such as derivatives or dubious mortgage securities. In the North Dakota system, the state collects taxes and fees from individuals and businesses. The state allocates those funds to state government agencies to cover the cost of running schools and other state services. The BND uses these funds to provide financial services to local community banks to help them, in turn, provide financial services to people and businesses in their communities.

The BND maintains about $4 billion in assets and has a loan portfolio of about $2.8 billion. The bulk of these loans are partnership loans with North Dakota's community banks. In a structure similar to the credit union business, the BND helps aggregate capital for about 90 to 100 community banks as they make mortgage and other loans to small businesses and farmers in their communities. The BND provides a secondary market to which local community banks can sell mortgages to shore up liquidity for more lending. The BND also underwrites commercial and state bonds for capital investments. In addition, it helps community banks manage their risk with loan guarantees to community banks that extend credit to local entrepreneurs. Though it is a for-profit enterprise, much of its profits are returned to the state's coffers in the form of rebates or dividends. The money earned in the state and taxes collected in the state are recycled in the state by a state-owned bank.

As I mentioned above, democratizing and re-localizing our financial sector will be an enormous undertaking. Much effort would have to be made to secure the capital for a local banking ecosystem. To help with this, the BND has a bank stock loan program specifically targeted for expanding its local ownership of banks. In this program, local investors who are interested in buying shares of a community bank in their area can receive financing from the BND.

Though it is operated as a for-profit institution, its programs and facilities are for the benefit of the state, the economy, and the people of North Dakota. Like a state-level Federal Reserve, the BND provides other services: check clearing, issuing currency, and direct deposit automated payment services. As the banker to the North Dakota state government, the BND can also help the state maintain stable funding to state agencies during times of budget shortfalls. Its CEO, Eric Hardmeyer, explains how it accomplished this and sacrificed some of its own profitability for the greater good,

> Back in 2001, 2002, when we went through the dot com bust, all the states suffered some sort of budget shortfall, including the state of North Dakota. At that time our budget shortfall was fairly insignificant—$40-some million. And so it was quite easy to overcome that. The governor just simply said alright, we're going to turn back 1 percent of all general fund agencies, and the Bank of North Dakota, you will declare another dividend to make up the balance.

In this way, BND functions as a stabilizing counterweight to the state that has to balance its budget during hard times.

The BND also has a program to help local municipal agencies. Unlike most city governments that finance capital development with municipal bonds, North

Dakota towns and cities can access credit through the BND. In this program, the BND, like any other bank, can borrow from the Federal Reserve's discount window and pass that money on to municipalities in the form of loans at an interest rate that is less than what they would pay in AAA bond markets. This would include emergency loans to help communities deal with wild fires, floods, and other natural disasters.

These financial systems—networks of credit unions, public banks, and community banks—stand as real world, positive institutional alternatives to our present dysfunctional and unstable system. Though in the case of North Dakota, much of the source funds originate in unsustainable industries like oil, coal, and industrial agriculture, yet the BND model can be seen as force for real change. Even if the public banks are siphoning money off of destructive conventional business sectors, it can recycle that money into sectors that are not destructive and are unconventional, and stay poised to carry the mantle of ecological permanence, vibrancy, and stability into the future.

A constellation of small scale financial institutions can be chartered to carry out the same mission as the Permaculture Credit Union or the Bank of North Dakota. When financial institutions like these form networks with other like-minded institutions, entirely new economic ecosystems can emerge. The bonds that tie these institutions together are like the strong attraction forces in nature that give rise to natural systems. To give a specific example of this, the Permaculture Credit Union formed such a bond with the Santa Fe Farmers Market Institute with the aim of securing land for local food production, which will be vital in the declining years of the Oil Age.

Feeding our people without the use of petrochemicals will perhaps be the most difficult challenge of the 21st century. Nonchemical based farming is arduous, labor intensive work. Farms in the areas surrounding Santa Fe are mostly small family operations with roughly three to nine acres of land per farm, and unlike large corporate farms, they cannot rely on the welfare of federal government subsidies. Like so many other small farming communities they cannot compete in wholesale food distribution markets that are dominated by the large corporate/industrial agribusinesses. And also like so many others, they organized farmers markets in order to sell directly to consumers— though even with the support of a farmers market, it is difficult to make a living and remain good stewards of the land. The majority of farmers in this region generate an annual income of $10,000 or less off their farms.

To make food matters more difficult, arable land in New Mexico is scarce and is systematically disappearing as a result of development, topsoil erosion, and damage caused by climate change. The region has lost over 200,000 acres of farmland and about 500 farms over the last decade, and fresh water is also in short supply. About half of it comes from reservoirs replenished from Rocky Mountain snow runoff and the other half comes from wells drilled on the land. Like everywhere else in the world, both sources are drying up.

With the coming end of the Oil Age, however, conditions for farmers are going to change. Chemical-based industrial agriculture will fade at a rate proportional to the decline of availability of oil. This will eventually pull farmers closer to their communities, food miles will necessarily shorten, and people will become more dependent on local farms. To plan for this future and to secure land in such a way that is healthy, functional, and supportive of small family farms, citizens and farmers in the Sante Fe area created the Santa Fe Farmers Market Institute (SFFMI).

The Santa Fe Farmers Market Institute is a nonprofit organization dedicated to securing a healthy and nutritious food supply for the local population and supporting local family farms. The Institute was formed in 2002 on a fifty-acre piece of land in what was once a rail yard and environmentally toxic brownfield acquired by the city of Santa Fe. Among other things the Institute has stitched together a network involving land trust organizations, local farmers and a farmers market, the State of New Mexico, the Permaculture Guild, and the Permaculture Credit Union to develop models for preserving farmland and for sustaining local food resilience.

One of their models is the Small Agricultural Land Conservation Initiative (SALCI). In this model, the State of New Mexico created a conservation easement program that incentivized the preservation of land for farming rather than allowing it to be used for development. As in most states, landowners in New Mexico pay property taxes on their land to the state. With this easement program, the owners can earn a tax credit reducing their tax liability up to a maximum of $250,000. In order to qualify for the credit, the easement must follow strict guidelines to protect wildlife, preserve scenic beauty, or to preserve open spaces for public benefit or agriculture. The SFFMI worked proactively to see that this easement program is used to preserve arable land for growing food. The process of putting land into an easement as such can be costly because the owners must pay for appraisals, mineral reports and other technical aspects of land use management. According to SFFMI, these fees can run up to as much as $15,000 which can be cost prohibitive for farmers who are trying to live on what already are poverty-level incomes. One way they invented to get around this problem is to sell the tax credit to investors for 85 cents on the dollar. This amounts to an advance on the tax credit at a 15 percent rate of interest.

At the same time, the SFFMI also opened a bridge loan program that they coordinate with Permaculture Credit Union and its affiliated organization, the Permaculture Guild. The program to help farmers get the funds up to $15,000 for the easement costs begins with tax-deductible donations to the SFFMI and the Permaculture Guild, both of which are nonprofits. Typically the SSFMI collects the first $10,000 in donations and the Guild collects the other $5,000. The total $15,000 goes into special accounts at the PCU where it earns a small amount of interest. The money in these accounts then stands as an unsecured line of credit for farmers. The farmers who qualify to participate in the program can tap into this line of credit to pay for easement costs as they arise. Once the tax credits come through to the farmer, they pay the loan back and the land is then protected and set aside for farming.

This program is just one of the many impressive and unconventional commitments the SFFMI has to food resilience and resourcefulness. They also offer educational services and workshops on cooking, nutrition, professional development, and microlending. The microloans, again, are coordinated with the PDU and can range from $250 to $10,000. All this work is structured around these institutional statements for their mission:

1. Assist farmers and other land based producers in the production and promotion of agricultural produce, products and value added products.
2. Promote, foster and encourage small farm and ranch operations and other rural land based operations in furtherance of the health, environmental, economic, social and cultural wellbeing of those dependent upon the land.
3. Engage in research, education, agricultural extension services, experiments, investigations, analyses and studies to benefit the advancement of agriculture and to foster and develop scientific methods for the application and dissemination of the results thereof.
4. Assist, promote, foster, encourage and preserve the historical land based lifestyle, traditions and culture of New Mexico.
5. Represent and advocate the legitimate common interests of farmers,

agriculturists, horticulturists, agro-artists, socially disadvantaged and other under-served land-based people.

6. Hold title to land and participate in the development of land in the furtherance of this mission.

7. Establish and maintain courses of study, educational activities and events related to the study of agriculture, value added agriculture and agro-art.

Each of these statements begins with an action verb signifying their proactive stance on creating local food resilience, overall economic wellbeing, cultural vibrancy, and social equity.

Along with the SFFMI is the Santa Fe Farmers Market, which is a separate for-profit organization that dates back to the 1960s. The market is owned and operated by the local family farms and artisans in the community and represents about 150 vendors. As an offshoot, the SFFMI was created to support the market and, among other things, to secure a permanent location and facilities for the vendors. The institute raised funds and built a complex including an impressive and aesthetic LEED-certified building that provides indoor and outdoor space for the vendors. They have the authority to sublease the office spaces to companies that pursue business practices consistent with the SFFMI mission. It has also leased space for a restaurant that buys food from the farmers market.

I had an opportunity to visit the market and was immediately struck by the festive atmosphere and spirit of active citizenry that imbued the place. People from the city browsed the outdoor stalls for fresh produce, handicrafts, jam, honey, and beautiful Bolga baskets hand woven by farmers in Ghana, Western Africa, who supplement their income by making and selling baskets. Aside from these baskets, the institute requires that the crafts must have 80 percent local content in materials. Inside the building are office spaces, a small cafe, a gift shop, and a 9,800 square foot hall where vendors offer their wares. The hall also moonlights as a banquet pavilion to be rented for events in the Santa Fe community.

The people at SFFMI and the Farmers Market have created more than just a place where buyers meet the sellers directly. Theirs is a treasure to be carried by the New Monastics into the future to emulate and build upon. The farmers market and the Institute have evolved something that has gone far beyond the dreams of Santa Fe's original farmers market pioneers of six families who set up "zucchini markets" in parking lots a half century ago. As envisioned by the meta-economists, they have created a vibrant sense of place that Randolph Bourne envisioned as the Beloved Community, a celebration of artisanal work alive with the intrinsic joys of work and sense of aesthetics close to the hearts of John Ruskin and William Morris, and local financing of small business operations and technology in the spirit of E. F. Schumacher. The growers and the consumers were brought together out of a sense of necessity, and what they have developed will continue to evolve as the world's supply of oil evaporates, as global warming causes the god of climate to rage, and as people quickly realize how important these institutional developments really are.

Through their hard work, they have cultivated a deeper sense of place, a celebration of the intrinsic joy in this work, all of which is imbued with the distinct coloring of local New Mexican culture. They exemplify what can be done in a community through a concerted effort to marshal whatever resources they have available to create new institutions that secure the needs of people—both the farmers and consumers of Santa Fe.

Local food resilience will continue to be a mounting concern for virtually every community on Earth as the Oil Age begins its descent. As our population expands, the majority of our additional numbers will be located in cities, and at the same time the decline of oil will push people to live in closer proximities. The need to sustain local food resilience in urban areas will intensify. The average food miles of the produce sold at Santa Fe's local farmers market is about forty-four miles, which is less than one-eighth of what the B Lab sets as its standard for local reliance in the supply chain. Such institutional developments will prove to be vitally important models as the Oil Age comes to closure.

The Urban Farmer and the Food Resilience Connection

Among the many sobering realities that peak oil expert Richard Heinberg shares with us in his book, *The End of Growth* (2011), is the precarious condition of the world's food supply. As if climate change, energy descent, and depletion of fresh water were not troubling enough, we are also being confronted with a more pernicious problem of peak phosphorus. Phosphorus is a key nutrient for sustaining plant growth and can be harnessed either through natural recycling and composting as is practiced in true organic agriculture, or it can be mined from phosphate rock and added to the soil and chemical fertilizers in conventional agriculture. Either way, phosphorus, like all minerals, is a limited and finite resource, and by itself will put limitations on plant growth including plants we eat or use for feedstock. Along with oil, copper, zinc, uranium, and everything else, phosphorus will be hitting its global peak and downward slide in the next twenty years or so. The grim result will be a decline in industrial agriculture, which is arguably the most inefficient way to farm ever invented on this planet.

Industrial agriculture is inefficient and unsustainable for a number of reasons. When you add up all the energy it takes to run tractors and other farm machinery and to synthesize fertilizers, herbicides, and pesticides from natural gas and oil, the energy input to food calorie output ratio varies anywhere between 10:1 to 300:1 depending on what you are producing. So when we start running down gas and oil supplies after we cross the peak threshold, food output will decline by a much larger proportion. This decline will be compounded by loss of soil fertility and higher toxicity caused by topsoil abuse wrought, again, by industrial agriculture. The story gets even more daunting when we consider the long distances food travels, energy intensive processing and drying, plastic packaging, and so on and disastrously so forth.

The implication is that conventional agriculture will eventually give way to unconventional food production. This reality combined with population densities arising from high transportation costs indicate that we will be relying more and more on food produced in the city or in our backyards. Small family farms and grocers were once an integral part of the social fabric in virtually every society around the globe. And although we cannot revert backward because evolution does not work that way, it appears that urban food production and consumption will once again become the nexus of social ties for future communities.

The idea of growing food in the city is not new. People living in cities have always grown and preserved food and composted and recycled waste for their own consumption or for the market. This intensive mode of food production has been practiced virtually in every culture in the world where there existed high concentrations of city dwellers and population densities, including cities in the US. During World Wars I and II, the governments of Britain and the US encouraged people to plant what were then known as Victory Gardens to alleviate some of the pressure on commercial food supplies brought about by the wars. People responded to their governments' calls and in the United States the gardens generated as much as 40 percent of the nation's food supply.

Today there are about 800 million people involved in urban farming world-wide. In the US urban farming, community gardens, community supported agriculture (CSA), and farmers markets have been growing rapidly, but they remain a very slim margin of the food business compared to dominant conventional and unsustainable food systems. This will change as industrial agriculture begins to fade.

Taking the lead in this transformation are SPIN Farmers® and urban homesteaders. The acronym SPIN means Small Plot Intensive farming and is a model for growing food within urban or suburban settings. It was started by a couple of urban farmers in Saskatchewan who developed a business model based on practical and easy to follow cultivation techniques with an emphasis on selling produce commercially through local markets, restaurants, and CSAs. The original SPIN concept was to engage in high value, intensive production of food that is rapidly perishable and not conducive to long distance transportation. The logic behind it was that these small farmers would not be able to compete with industrial agriculture that can truck food that is slow to perish over long distances. The farms are small franchises typically located in sub-acre plots like people's backyards or vacant lots. All the work is done by hand and in some examples the produce and composted inputs are transported around the city with pedal power—bicycles equipped with cargo trailers that can carry as much as 500 pounds of freight. The salient characteristic of SPIN farms is that they are extremely high yield and generate much more food output as a ratio to inputs compared to conventional agriculture.

One of the greatest advantages of SPIN farming is that the plots are small in scale compared to large commercial farms and they require relatively little capital investment to get started. Startup costs can run between $10,000 and $25,000 depending on what size farm, the quality of the soil, and other factors. It might prove difficult for prospective SPIN farmers to get bank loans as the business model is unconventional, though this is the kind of project that could be financed by a local credit union or community bank. The projects are light enough on technical requirements so that they are relatively easy for a novice to learn. Part of the reason why this model requires little startup capital is that the cultivation is done primarily on other people's land. SPIN farmers found that there are a fairly large number of homeowners who have grown tired of trying to maintain highly manicured lawns. They welcome the SPIN farmers to come to their yards and work the property into clean and aesthetic gardens. Also the owners can get a share of the produce and possibly a share of the revenue depending on the arrangement. Zoning can be a problem if the city has laws against commercial farming in residential areas, though as we will see in examples in the Pacific Northwest, cities are amending their zoning laws to gear up for the coming wave of urban farming.

Noncommercial urban homesteaders also practice intensive food production in metropolitan areas. Unlike SPIN farming which is deliberately commercial, urban homesteading is about food self-sufficiency for individual households in the city. Homesteaders are generally passionate about growing their own food anywhere it will grow: backyards, rooftops, indoors, window boxes, or alleyways. Homesteading is a do-it-yourself paradigm that encourages people not only to grow their own food, but to build things to be used in urban farms like water catchment systems, grey water processing systems, compost toilets, and small-scale renewable energy.

Whatever model is used, urban farming is not just about growing food. As an ecological movement it involves virtually no food miles and no requirement to use oil or natural gas-based chemicals. As a social movement it is about drawing members of a community together around food production and consumption and filling the void between growers and consumers in conventional systems. People naturally want to see where their food comes from, how it was grown, and who grows it. They want it to be fresh and aesthetically pleasing. Communities that bond in this way rejoice around harvest times with festivals and celebrations and support each other when harvests are not so bountiful.

There are sideline issues with urban food production and they will have to be governed institutionally with sound rules and shared strategies. For one, city dwellers will be concerned about how the livability of their neighborhoods will be impacted when the places where they live become centers for re-localized economic activity. Good city planning and zoning ordinances will be crucial to managing the small cottage industries that will spring up in residential areas. What might seem like vibrancy to one household might be considered by another to be noisiness, smells, and generally worsening conditions due to crowding. It would be reasonable for residents not to want truckloads of horse manure and rotting cabbage piling up in the yard adjacent to them. The intensity of SPIN farming requires enormous amounts of composting to maintain the yields the farmers seek. This in turn requires that farmers spend much of their time seeking sources of compost generated in places outside their planting areas. As these inputs are drawn from resources in other places this raises a potential concern about the sustainability of intensive urban farming. It is possible that, just like conventional agriculture will be hitting the wall of peak phosphorus and chemical fertilizers, urban farms may hit the wall of peak compost, bone meal, and other resources needed to sustain them. The same is true of aquaponic systems that require expensive greenhouse structures and a steady supply of fish for recycling nutrients and generating fertilizers in the water. And of course, as in all farming systems that have ever existed in the history of the world, there will be competing interests over how the city's water supplies will be allocated in the midst of the heightened demand for what will certainly be declining supplies as conditions of global warming intensify.

Food can be a bond that holds people in a community together like nothing else. Models of urban food production are treasures to be carried and modified by the New Monastics. They offer a tremendous opportunity to develop local sources of healthy food and help create vibrant and sustainable communities. Secondary production systems can also spring up from urban farms such as local compost exchanges, valued added products such as canned food, dressings, or sauces, producing transport equipment, storage facilities, greenhouse projects, and so on. These are just some of the ways economic re-localization can evolve, but for this to be effective there must be good planning and institutional control by local government.

The Local Government Connection

In my research I have spoken with many business people about positive change. Their motives for wanting to have a positive impact on our world is clear and openly stated, but they are compelled by the notion that initiatives for positive change are most effectively carried out in an open market environment. There is a virtual consensus among them that their commitments to the ecological and aesthetic qualities of clean renewable energy, the art of making beautiful objects, generating clean energy, preserving ancient traditions of craft brewing and farming, and a general sense of wellbeing and ecological permanence cannot be legislated by government policies.

There is much truth in what they say. But we have to be careful about this because this could easily be interpreted as an endorsement of neoliberalism and the disasters it brought us in recent decades. Generally speaking, entrepreneurs chafe at the notion that the local government will impose new rules on their activities, yet this old "us versus them" attitude toward government is an impediment to positive change. It may be true that you cannot legislate social or ecological consciousness, but it is equally true that you cannot buy it in the open market. All too often people try to dichotomize our options for a commonwealth as either "free market capitalism" or "state socialism." But no economy in the world falls purely into either of those categories. Rather they are pluralistic or mixed systems. Values-based government legislation and genuine democratic reforms are just as important as values-based entrepreneurship in creating real and lasting change. Business models do not stand alone. They work and succeed because they are embedded within a broader structure of institutions and culture that are at least partially shaped by governments. Our social consciousness is part of the meta-economic core that Schumacher refers to as the "eye of the heart" and which transcends both legislative processes and market forces. As we cultivate this heart with wisdom, compassion, and mindfulness, we can build new institutions of all kinds.

In their book, *Be The Change* (2009), authors Thomas Linzey and Anneke Campbell tell the stories of active citizens who ". . . have transformed the members of local governments from mere administrators into the lead wave of a movement toward sustainability through local self-government." If we step back and look at the bigger picture, which is what a commonwealth is, we see that these entrepreneurs owe their success, at least in part, to government ordinances on land use, statutes and regulations on banking, public utilities, and state and local subsidies. They operate in a market environment, but that environment is embedded within a broader institutional structure of rules, norms, shared strategies—institutions.

None of us can claim to have invented the concepts of ecological permanence, artisanal traditions, or social justice. We have inherited these concepts from those who lived before us and we are contributing to their evolution for those who will follow by changing our institutional structure in a direct and mindful way whether they are in the public or private sectors. Here we find more treasures for the New Monastics in the public sphere, particularly with local government.

It is easy for people to become cynical about government. Most of us don't enjoy having to pay taxes and it seems that the only things that score media attention are scandals, pork barrel corruption, and government waste of taxpayers' money. Nearly all of the great meta-economists we have explored from the transitioneers to Schumacher recognized the importance of an active citizenry for projects of ecological permanence, civic regeneration, and for the empowerment of the individual.

The emphasis here is on local self-government of an informed citizenry, but that is not to say that national politics is completely irrelevant. Mumford and the Young Americans were wise to be critical of an overemphasis on national statecraft, though it would be to our detriment to ignore national politics altogether as the serious economic and environmental problems we are facing are affecting all of humanity, and effective action will require some level of national or global consensus. But rather than passively waiting for answers to come from top down reform legislation, citizens can push up with serious action at the local or state level. An active and empowered local citizenry can be very effective in securing the rights of individuals and communities and protection of natural habitats. Each community that is politically energized will join with other communities to form a foundation or platform for broader political and economic transformation. And the more effective they are at this, the more leverage they will have in pressing for better national legislation for coping with the end of the Oil Age.

As energy becomes scarcer, pressures will intensify to mine for coal, drill for oil, dig energy out of tar sands and shale, and frack natural gas. The main source of this pressure is the energy industry that sells energy to the consuming public. But the industry has a fairly easy time enlisting local or state government as well as the general public to their cause. Their seduction is familiar and basic—we need to come in and extract this energy to avoid shortages and also to generate revenue for schools and create jobs. Though sometimes this seduction is based on false promises, it is also often the truth. Mining and drilling can create boom towns and money. The mineral resources in these boom towns, however, eventually are exhausted and the ephemeral boom turns to bust. So it would appear that local citizens have to make some hard choices. Do they sell out to the energy companies for short term gain or do they commit to the long term work of building a new and different economy that is not so dependent on high levels of energy consumption and the energy industry for livelihoods?

The Community Environment Legal Defense Fund (CELDF) is helping those in communities who are willing to fight the long, hard battles for building a new community. Their focus has been specifically on communities that are struggling against large energy companies that use their muscle and political clout to take over and plunder the land. CELDF has developed a model to assist community groups and local government in drafting new laws that protect the rights of communities and their habitats from powerful companies. A first step in their model of community activism is to establish a vision of what kind of community and ecology they want to live in, and to ask who in the community is going to be making the decisions that will shape into this vision. Once their vision is clear, they move forward to adopt a home rule charter that codifies this vision into local law—genuine institutional transformation.

One such community with a home rule charter is Blaine Township, Pennsylvania. Blaine is a small town about forty miles outside Pittsburgh in Western Pennsylvania. The economies of most towns in this area are dominated by the resource extraction industries and the coal companies own most of the mineral rights throughout the region. For over two centuries, coal mining had not reached Blaine, but that was about to change as the country's largest underground coal mining company, Consol Energy, began to covet the coal lying in the ground beneath the homes, shops, and farms of the Blaine area where the company has claim to mineral rights. This claim is protected by old state laws which give companies like Consol much leeway to control the economic affairs of local communities. But mineral rights and the laws that protect them are not shackles and chains that hold people in bondage. Citizens can make new rules to strip corporations of their institutional powers with as much force as the corporations themselves throw around like a blunt weapon. In Article I of Pennsylvania's Constitution is the following institutional statement:

> All power is inherent in the people and all free governments are founded on their authority and instituted for their peace, safety, and happiness. For the advancement of these ends they have at all times an inalienable and indefeasible right to alter, reform or abolish their government in such a manner as they think proper.

The people of Blaine take these powers seriously. If there were ever a juridical justification for challenging laws that protect mineral rights, it would be that the mining processes themselves are ecological catastrophes.

The method used by coal mining companies in this area is known as "longwall" mining. This is basically a process of using giant mole-like machines that tunnel several hundred feet underground, grinding and chewing everything in their paths into bits to be hauled to the surface and combed for useable coal. The process causes the ground surface to sink, destroying watersheds, aquifers, and other groundwater sources. Wherever longwall mining occurs it is turning everything underground into sludge and mud. That along with drilling and fracking for natural gas is contaminating the water supply and people are forced to store water above ground in 550 gallon plastic tanks known as "water buffaloes." The water in the tanks is shipped in from external sources and the tanks sit in peoples' yards as an eerie reminder to the residents of the damage being done underneath their feet.

Between 2006 and 2008 local residents in Blaine took preemptive action with the help of CELDF. Local activists took legal action and passed a series of city ordinances banning the practice of longwall mining within the township. The ordinances, among other things, claim that communities have the constitutional right to control business activities within their jurisdiction. The ordinances ban coal mining and require businesses to publicly disclose all of their activities and, perhaps most significantly, they strip the claim that corporations are "persons" and in the letter of the law it states: "This illegitimate bestowal of civil and political rights upon corporations prevents the administration of laws within Blaine Township and usurps basic human and constitutional rights guaranteed by the people of Blaine Township."

Pennsylvania, like most states, has laws that allow cities or counties to adopt their own governing constitutions known as home rule. This shifts the power of local governance away from the state to the local community. The charter can be drafted in such a way that will allow the citizens in the community to establish a body of laws which can regulate business activities in accordance with social and environmental values. Contained in the Blaine charter are these institutional statements:

Section 304 *No Institutional Rights* Government, human institutions, corporations and other agencies of society are servants of the People and shall not have rights, but are obliged to remain subordinate to the People and to their rights.

Section 320 *Right to a Healthy Environment* All residents of Blaine Township possess a fundamental and inalienable right to a healthy environment, which includes the right to unpolluted air, water, soil, flora, and fauna, and the right to protect the rights of natural communities and ecosystems, upon which each resident is both intrinsically a part and dependent.

Section 324 *Rights of Natural Communities* Natural communities and ecosystems, including, but not limited to, wetlands, streams, rivers, aquifers, clouds, and other water systems, possess inalienable and fundamental rights to exist, flourish and naturally evolve . . .

Section 325 *Life is not Property* Genetic characteristics shall never be property or be altered to become or be claimed as property, and no person, corporation or institution shall claim ownership of the genetic characteristics of any living being or organic tissue.

Section 326 *Right to Water* All residents, natural communities and ecosystems in Blaine Township possess a fundamental and inalienable right to sustainably access, use, consume, and preserve water drawn from natural water cycles and sources that provide water necessary to sustain life within the Township.

Section 327 *Right to Establish Sustainability Policies* The People of Blaine Township retain the right to establish and promote sustainable policies through their Township government on issues including but not limited to energy use and production, water, waste, environmental preservation, land use, and all issues relating to the quality of life for People and Nature.

Section 330 *Corporations Subordinate to the People* Corporations are human institutions, created by the People through their government. The People and their rights shall never be subordinate to corporations . . . The conferral of constitutional and legal powers upon corporations by the legislature and by the courts . . . shall not be recognized as legitimate by the People of Blaine Township or their government.

Section 331 *Privileges of Corporations Not Recognized* To ensure that the privileges of a few never subordinate the rights of the People to make self-governing decisions, within Blaine Township corporations shall not be "persons" under the United States or Pennsylvania Constitutions, or under the laws of the United States, Pennsylvania, or Blaine Township, and so shall not have the rights of persons under those constitutions and laws.

There is no doubt that these reform initiatives will be challenged in court. Such cases could end up in the US Supreme Court and could potentially alter the structure of power as it now stands. It wouldn't matter whether the higher courts uphold these ordinances or not. The burden would be on the companies to appeal and such legal appeals are costly. This process of waging guerilla legal battles will fatigue the businesses that resist the values supporting these community initiatives. And what is more important than the outcome of any particular case is the momentum that these cases can generate. Ironically, we can learn much from the efforts of the lawyers who pushed for corporate personhood in the 19th century. In the early stages of their movement, the courts ruled against them over and over again until they got a lucky break. Abolition of slavery went through the same process, as did the labor movement, women's movement, and civil rights. In each instance those who were pushing for institutional change had to endure one defeat after another before their aims were realized.

If, at the end of the day, the citizens' aims are realized, what do they do then? It is easy to villainize large coal companies as they are involved in the dirty and destructive business of extracting energy. Yet Americans burn 2.7 million tons of coal and 63 billion cubic feet of natural gas every day. If we seek to protect ourselves from the destructive practices of energy companies, we need to also be working toward protecting ourselves from ourselves. To do that, however, would require a very distinct and different way of living in the world. If people in these communities do not build new institutions that provide them with the means to sustain their livelihoods without the big energy companies, then their communities will suffer the same fate as the Reconstruction effort of the 19th century—the replacement of one catastrophic system for another. If we hold onto established or conventional practices that are, in a broader sense, anchored to growth and accumulation, then our efforts to achieve ecological permanence are certain to fail.

As a general rule, governments tend to be reactive rather than proactive. They will eventually come around to deal with messes after they've been pressured by the citizenry to the point of intolerability. Some local governments, however, are proactively seeking the kinds of changes necessary to adapt to energy and food scarcity. For example, in the city of Seattle, Washington's Department of Planning and Development is drafting code changes not only to support urban farming, but to encourage it. As urban agriculture can potentially be a messy affair, sound urban planning and policy is a must. One of

Seattle's first initiatives is to clarify the terminology; that is, to make precise definitions of what is meant by "horticulture," "aquaculture," "animal husbandry," "community gardens," and "urban farms." It is also proposing to allow community gardens and urban farms in all zoned areas: industrial, commercial, and residential, though with square footage restrictions such as limiting urban farms to a maximum of 4,000 square feet per area of planting. In residential areas the city will increase the number of chickens allowed to be raised from three to eight. Seattle is also planning to allow downtown residents to build rooftop greenhouses, though subject to size restrictions.

Another forward-looking city in the Pacific Northwest is Portland, Oregon. Portland is distinguished for its progressive urbanism, its obsession with recycling and sustainability, light rail transit systems, and the legendary independent Powell's Bookstore, as well as its gray skies and drizzle. Like other places in the Pacific Northwest, Portland, the City of Roses, is also noted for its stunning scenic beauty. The city has been named one of the most vegetarian friendly cities in the US as well as among the most walkable and bicycle friendly cities in the world. Portland is also globally renowned for being one of the most well prepared major cities in the world for livability at the end of the Oil Age. To make a point of disclosure, I am a long-term resident of Portland, but this means that I have also been scratched by its rougher edges, which are numerous.

The effort to preserve the scenic beauty of Oregon goes back forty or more years. In 1973, under the leadership of former Governor Tom McCall, the state passed a series of statewide land use planning laws. One of them restricted urban and suburban sprawl by placing strict urban growth boundaries (UGB) around its cities and this was most stringently applied in the Portland metropolitan area. Oregonians decidedly resolved not to replicate the seemingly endless sprawl and freeway expansion that is characteristic of other cities like Los Angeles or Houston with their long car commutes and foul smog. Instead Oregon opted for density by keeping development contained, which shortened distances between homes and workplaces, lessened the city's reliance on fossil fuels, and served to protect local farmland and watersheds.

The urban growth boundary for Portland was wrapped around the city in 1979. Though some provisions have been made to accommodate population increases, these provisions are very limited. Since Oregon's land use laws were passed in the early 1970s, Portland's population has increased about 60 percent but has only allowed the land area to expand by about 2 percent.

Throughout the last three decades Portland's model was harshly criticized for being wrapped too tight, particularly from anti-government neoliberalists who contend that by not allowing developers to buy local farmland for sprawl and subdivision, housing shortages would emerge and become chronic problems making the areas within the boundary unaffordable. Nonetheless, there remains no evidence of this claim as Portland remains more affordable than all the other major West Coast cities. Moreover, the land use laws require that the UGB be modified every twenty years or so to deal with population pressures.

As the city was preparing for density with the adoption of the UGB, it abandoned previously developed plans to expand its urban to suburban freeway system. As oil becomes so expensive that car commutes begin to fade from viability, transportation planning is vital. Portland and the surrounding counties reallocated the funds that were to be used for freeway expansion to build a light rail transit system that would serve the same transportation needs.

One of the goals was to simultaneously deal with heavy freeway traffic and comply with the Environmental Protection Agency's (EPA) Clean Air Act (1972) of which the city was in constant violation. The fifteen-mile rail line was completed in 1986. The rail line was named MAX, meaning Metro Areas Express, and despite the shrill criticisms from conservative neoliberals and not-in-my-backyard types, it proved to be a major success and evolved to become a model for good transportation planning.

The Max system minimizes the impact on existing infrastructure by using freeway right of ways, shoulders, abandoned freight train lines, and medians that run the trains down the middle of boulevards. With the momentum gained from the light rail system, Portland opened a streetcar line in its downtown area. The streetcar is operated by Portland Streetcar, Inc., a nonprofit, public benefit corporation. All of the transit stations for Max and the streetcars are linked to bus lines, and in the early 1990s Portland Metro Area planning authorities, in conjunction with the plans to expand MAX to the metro areas suburbs to the west, implemented their "2040 Plan."

As the name would suggest, the 2040 Plan was looking at transportation issues decades into a future characterized by scarce and expensive fossil fuels. In a collaborative effort among the city, three counties, and several suburbs, the local governments controlled development to become "station communities" clustered around MAX stations yet containing affordable housing, shops, and other necessary amenities. In other words, transit stops were not just places where people waited for the train, they became mixed use, pedestrian oriented community centers. In addition, the plan included programs to focus other development projects to infill existing vacant areas located within the UGB. The planners' ultimate goal has been to absorb what they project to be a 1 million plus increase in inhabitants over the following decades without significant increases in sprawl, freeway congestion, or greenhouse gas emissions. At the time that seemed like a utopian dream, but the project has since gained much momentum and is well on its way of achieving its goal. It has also inspired the city planners and officials to do more.

Portland became the first major city in the US to enact a comprehensive plan to reduce CO_2 emissions. In 2009, the Portland Bureau of Planning and Sustainability launched a new plan to deal with climate change and once again Portland was put on the global map for progressive ecological planning. This was not Portland's first response to climate change. In 1993 it drafted a "Carbon Dioxide Reduction Strategy" and in 2001 it unveiled a joint county-city "Local Action Plan on Global Warming" that became a source of inspiration for the founders of the Transition movement. The latest efforts, The Climate Action Plan of 2009 and The Portland Plan of 2012 are far more rigorous and comprehensive, though not without flaws.

The Climate Action Plan (CAP) was a response to a resolution adopted four years earlier to design a strategy to reduce carbon dioxide emissions by 80 percent by 2050 and in the interim set a goal of 40 percent reductions by 2030. Unlike many other cities Portland is blessed with an abundance of low carbon hydropower, though it still generates about 68 percent of its overall electricity from coal, natural gas, and nuclear power. One of the key provisions in the plan was to retrofit Portland's existing public buildings for energy efficiency and to assure that any new building in the city must be made with maximum energy efficiency. In partnership with other institutions, Portland has also created an energy trust program that provides low interest financing for homeowners to make improvements for energy efficiency. The interest payments from the financing are factored into the monthly bills amortized over fifteen or twenty years. In its continued effort to promote density over sprawl, Portland planners also consciously zoned housing development projects to create vibrant urban communities and diversity of income in each neighborhood. Zoning laws required that buildings have street-level windows and sidewalk facilities to create a more lively community atmosphere, affordable housing must be built along with larger upscale homes, and everywhere the city has green spaces, fountains, and open spaces to create a feeling that some have described as "Portland is not a city that contains parks, it is a park that contains a city."

Portland has found itself in the limelight as a model that urban planning expert Carl Abott and others have described as "the capitol of good planning." James Kunstler thought of Portland while reflecting on the ideals of Lewis Mumford and asserted that Portland's unique obligation to both environmentalism, aesthetics, and urbanism ". . . was Lewis Mumford's dream come true: authentic regional planning." Joining the chorus of praise, *Grist* magazine named Portland as the second most eco-friendly city in the world behind Reykjavík, Iceland:

> The City of Roses approach to urban planning and outdoor spaces has often earned it a spot on the lists of the greenest places to live. Portland is the first city to enact a comprehensive plan to reduce CO_2 emissions and has aggressively pushed green building initiatives. It also runs a comprehensive system of light rail, buses, and bike lanes to help keep cars off the road, and it boasts 92,000 acres of green space and more than 74 miles of hiking, running, and biking trails.[55]

It remains unclear where the financing for all of its ambitious greened up infrastructure is going to come from. Portland is not an affluent city and it may very well see source funds evaporate like what seems to be happening everywhere else in the world. It should also be noted that Portland's commitment to social equity is tepid and conventionally boring. Social equity remains captured by the worn-out cliché of triple bottom line business growth. As with other conventional models, Portland hopes to grow its way out of its social problems and in a bizarre turn of mind thinks that overseas exports are the key—exports requiring enormous consumption of fossil fuels that cannot be sustained.

Portland's unemployment rate is among the highest in the country. Extending from an extremely outdated system of public finance, Oregon's school system remains chronically underfunded and dysfunctional. Only 53 percent of its high school students graduate in four years and schools currently have a dropout rate of 23 percent. One quarter of the city's population are working poor and unable to meet their basic needs because of lack of income. Many of its density projects have backfired in gentrification of some of Portland's inner city neighborhoods, further marginalizing its already underserved population to make room for what many have perceived as an influx of the white and educated who can drift with leisure on inherited trust funds.

Nonetheless, Portland has been working toward easing the transition to a world without oil in ways that perhaps set it far ahead of many other cities in the looming peak oil timeline. The most promising element in the Portland Plan may be its Healthy Connected City models. The goal of these models is to create a system of neighborhood hubs linked by a network of greenways that integrate nature with each neighborhood. Echoing Kunstler's reflection, this is a plan that would make the dreams of not only Mumford, but also Patrick Geddes and Ebenezer Howard come true. Building on the 2040 Plan for the broader metro area, the Portland Plan seeks to build neighborhood hubs that will create vibrant community centers with all the necessary amenities for living within a twenty minute walking radius. Neighborhood greenways will include biking and walking trails, storm water collection systems, native fauna and flora habitat, and urban forestry, all within the city. The neighborhood hubs are to be connected by civic corridors that link the hubs together with streets and transit arteries moving with low-energy-intensive transportation systems.

As with Ebenezer Howard's project on Garden Cities, critics are quick to bash away at planning or vision as being utopian, dreamlike, or idealistic—as if those were inherently bad things. Granted, if a plan has no hope of becoming a reality then it is probably not a good plan. But decades ago critics tried to tear down the plan to build Portland's MAX system because they didn't believe people would ride it, and now it has become a model of public transportation planning. In their newly revised book, *Transportation Revolutions: Moving People and Freight Without Oil* (2010), authors Richard Gilbert and Anthony Perl write:

> High oil prices will cause changes, but the change will be destructive if not anticipated, as occurred in 2008. In one of many dismal scenarios, car-dependent suburban residents who can no longer afford to refuel their cars, and have no alternative means to travel to work or buy essential goods, will have to abandon their homes or live at a subsistence level on what they can produce from their backyards. If a region dependent on food imports by truck can no longer afford the transport costs, and there are not alternative means of moving food, residents will have to rely on what can be produced in the region, which may be too little for the numbers of people who live there.[56]

Reflecting on this passage, one has to wonder why it is so easy for the cynics and critics to accept such a dismal scenario, particularly when dreams of the past have come true in ways that have far surpassed the visions of the dreamers.

The process of crafting new institutions at the end of the Oil Age will necessarily involve political action at the local level. It is becoming increasingly clear that we cannot wait for others to step in and solve problems for us. Evolving a new commonwealth is not going to be successful if we expect it to be a top down effort sponsored by federal reform legislation. Reversing the order of decision making and governance from corporate lobbyists in Washington, D.C. to citizens will be most effective if it is initiated by activists in cities and communities around the country. As an active citizenry pushes upward against the recalcitrant corporate, state, and federal institutions, they can favorably alter the structure of power. But this can only happen if people make a concerted effort to rewrite local laws to make government agencies and businesses in their communities democratically accountable. Even the most progressive and forward looking cities like Portland need to be pressured by their citizens to do what is best for the common good. With time and with a vision inspired by the meta-economists, citizens and local government can rewrite their rules and renew their shared strategies to make this vision a reality. To do this, they will also need progressi[...] entrepreneurs, who, like the founders of the B Lab, want to transform [...] te sector with the same goals, values and vision. For this effort we [...] ne treasures I found in Colorado.

The Colorado C[...]

Creating a vibrant re-localized commonw[...] about coping with energy scarcity. It is about coping with en[...] ys that are socially equitable and infused with a sense of plac[...] joy of good work and craft traditions. A key aspect of this [...] ligned with these values while economically enfranchisin[...] community with ownership. As we develop new business [...] we get closer to realizing the kind of commonwealth env[...] -economists. I found some examples worthy of followi[...] untain state of Colorado.

One of these is Namasté Solar, a certifie[...] naste specializes in building custom, renewable energy g[...] tion systems.

Their primary business is the design and installation of solar energy systems for commercial and residential use. Namasté was created in 2005 as a result of a combination of local entrepreneurship and citizen activism. Voters in Colorado had passed a state initiative requiring that the state's largest utilities obtain 3 percent of their electricity from renewable sources by 2007 and 10 percent by 2010. This was the first legislation of this type of in US history, and it created an institutional and business environment more conducive to the use of alternative energy.

The name of the company, Namasté, is a Sanskrit greeting or salutation. The founders chose the name in part because one of the founders had just returned from a three year stint working in solar energy in Nepal. It was also chosen because the original meaning was to recognize someone's individual existence and at the same time that this existence has a deeper connection to everything else in the universe. As with the scientific movements of the 1970s that gave rise to design system of permaculture, this recognition signifies a holistic or ecological frame of reference for business and economics. This holistic view of Namasté's founders helped shape their business model to bring renewable energy to households and businesses in a way that is fair and socially equitable; that is, to recognize the importance of each individual's stake in the community. To that end, Namasté is structured around a nonhierarchical and democratic model of governance and co-ownership.

Namasté is 100 percent employee owned through a customized shareholder agreement. Though it is owned by employees/co-owners, not all the employees are co-owners. As of 2010, the company employed seventy people and over half of them are co-owners. Like that of most corporations, Namasté's governance rests with the sovereignty of the shareholders and board of directors. Unlike most conventional for-profit corporations, however, their decision making process is uniquely and intrinsically democratic. Individuals are encouraged to take responsibility and make their own decisions relevant to their competencies, yet do so with company support. Broader company-wide decisions are made by committees or teams, although, if a matter is considered too important to leave to a committee, the board of directors can intervene or veto any decision and instead take the issue to a company-wide vote. Ultimately the decision making process can end up requiring a full vote by shareholders who are also the employees. As the company is built around these institutional rules, it maintains a tightly democratic business culture.

Part of their business culture is rooted in the Buddhist concept of "right speech" or what they call the FOH model of communication—Frank, Open, and Honest—but with an emphasis on positive action rather than gossip or intrigue. The people at Namasté have worked hard to create the right business ambience conducive to productivity and the joy of work, sustainability, and a broader sense of interconnectedness with their community. This is reflected in their mission statement to ". . . work in Colorado to propagate the responsible use of solar energy, pioneer conscientious business practices, and create holistic wealth for our community." Namasté's mission differs from a typical capitalist enterprise that is driven to accumulate wealth for wealth's sake. Holistic wealth in this context means creating a wealth of benefit that is not limited to a narrow measure of financial success for the owners, but for all stakeholders in the community including their customers and the fauna and flora of their surrounding habitat. This mission, institutionalized in their guiding principles and founding documents, is protected by what they call the "walls of the castle" that safeguard their mission as the company remains a going concern in the community while personnel come and go over time.

Namasté maintains healthy ties with its community in other ways besides creating holistic wealth and clean energy. They have initiated programs for educating their community on issues related to renewable energy and donate 1 percent of their revenues back to their community for grants and sponsorships to help the community deal with problems of homelessness and to raise environmental awareness and support local culture. And since its inception it has donated over 150 kilowatts of solar energy construction projects as part of their in-kind giving program to community organizations.

Its commitment to clean energy, democratic governance, and holistic wealth has earned Namasté many accolades. It was named WorldBlu's "Most Democratic Workplace" in 2010 and 2011, received Boulder's Chamber of Commerce "Best Place to Work Award" in 2009, and also in 2009 was named by the *Denver Business Journal* as the "Best Place to Work." In the spirit of Ruskin and Morris, Namasté's commitment to energy efficiency is, in its own right, a commitment to aesthetic form. This was stated succinctly by one of the co-owners: "What I like about renewable energy is how elegant it is. There are no smokestacks; no stinky and noisy generators. Waste is ugly. It's gross."

There remain challenges for Namasté and the solar industry in general. Like all renewable energy infrastructure, capturing and harnessing energy from the sun requires significant capital investments. Up to now, the burden of aggregating that capital has been lightened somewhat with state and federal subsidies. Namasté's business, measured in terms of revenues, has grown by 2,500 percent since it began. Much of that growth has been stimulated by about 50 to 60 percent cost subsidization from government. These subsidies are temporary. The idea is to have the government support the industry in the short run so that it can grow and develop to a point where it reaches economies of scale and benefits from technological progress that will help bring the costs down. And costs have come down as a result. Eventually, though, the government support will be withdrawn with the expectation that the industry can stand on its own. Once this happens, the fate of the industry will remain to be seen.

The solar industry faces other challenges. US producers of solar panels have been clobbered by low cost imports coming from China. Some mainstream environmental organizations, however, applaud the cheap imports, stressing that affordability of the panels is critical for us to kick the fossil habit. Companies that install solar systems, like Namasté, raise questions about the quality of the panels coming from offshore as well as the ecological conditions under which they are manufactured. Nonetheless, with the coming end of the Oil Age we will need to harness whatever renewable energy we can get our hands on. For that reason, solar will remain a solid and viable industry and a key aspect of our energy systems of the future. Nevertheless, as I have stressed before, we simply won't be able to use anywhere close to the amount of energy we consume with fossils.

One of Namasté's commercial installation projects was an impressive array of panels built on top of New Belgium Brewery about sixty miles to the north of Boulder in Fort Collins. New Belgium Brewery (NBB) was established in 1991 after one of the founders returned from a bicycle trip in Europe. While he was in Europe, his idea for a business germinated from two fascinations: the Belgian style of craft beers and the classic cruiser style bicycles with big rubber tires. Back home in Colorado, he and his partner acquired some repurposed dairy equipment, some wooden barrels, and a home brew kit and started making beer in their basement. They carefully followed the Belgian tradition, which in the European beer-brewing culture means a bold deviation from orthodox tradition. That is to say, it means exploring new frontiers with creative innovations to the recipe. One of their first creations was a brown ale they named "Abbey" as it is a descendent of the Belgian Abbey ales that can be traced back to the 19th century version brewed by a Trappist monastery in northern Belgium. Another is an amber called "Fat Tire" that, once they went commercial with their brewery, became their flagship product and became a symbol for their enthusiasm for bicycling.

New Belgium is now the third largest craft beer brewery in the US, but it is not just about brewing beer, it is also a technological, ecological, and cultural phenomenon. The main building is exquisitely renovated from an old sugar processing plant. As one visits the brewery, one is immediately struck by the bright and lively atmosphere and a celebration of creativity and artisanal crafts. The original dairy equipment and wooden barrels are on display like museum pieces and have been replaced by state-of-the-art brewing and bottling equipment with the capacity to crank out seven hundred bottles per minute. And next to the old dairy equipment is a small collection of antique cruiser bicycles.

Anyone familiar with the New Belgium brands cannot help noticing a plethora of images of the fat tire cruiser bicycle, which eventually became their company logo. Since 1999, the brewery has commissioned bicycle manufacturers such as Felt, Schwinn, and Electra to custom produce vintage models of cruiser style bicycles exclusively for their employees and charity raffles. As part of the employee ownership model, once coworkers become vested after the first year they receive one of these gorgeous vintage style bikes. After three years they get a free trip to Belgium. Aside from being advocates of the obvious health and ecological benefits of bicycling, the company supports the skilled craft of rebicycling. Throughout the brewery are bicycles and things made from old bicycle parts. In the tasting room, beer drinkers sit on slick barstools made from old rims and frames that might otherwise have ended up in a landfill, and on the wall are bottle openers and other novelties made out of sprockets, chains, and other bicycle parts.

These craft items were made by a small company located in the scenic Columbia Gorge area of northern Oregon called Resource Revival. According to one of the owners, "Creating a free, demand-driven recycling service for hundreds of bike shops that also provides jobs for ten employees, supports dozens of suppliers, etc., and—perhaps most importantly—makes people say 'wow' when they see a bowl or bottle opener made out of bike chain. What's the Net Present Value of a smile?" This is ecologically sound work that is done in the true artistic spirit of John Ruskin who believed that such work leads toward the full development of a human being.

New Belgium sponsors their famous Tour de Fat, which is a kind of traveling circus featuring a bike rodeo, live music and other stage performances, and a colorful bicycle parade. The Tour also evangelizes swapping cars for bicycles and the use of bicycle trailers, all in a tremendous carnival of celebration. Tour de Fat moves its circus around the country by truck and trailer and on the sides of the trailers are painted images of people riding bicycles with wind turbines in the background.

The company offers other treasures beyond these celebrations of creativity and craft traditions. Like Namasté Solar, it is firmly integrated into the local community through ownership. The business is 100 percent internally owned; the employees, or coworkers, own about 43 percent of the company and the balance is held by the original founders. As coworkers/owners, the employees have a high level of engagement in the operations of the company. This includes sharing responsibility and accountability as well as being able to take credit for what they all do well. Moreover, their commitment to the enfranchisement of their workers stems from an underlying ethic of social justice and equity, and this easily overlaps with a strong environmental ethic. This core ethic, which has the status of an institutional statement, is expressed as, "honor nature at every turn of the business"—a rule to live by.

Early on New Belgium took measures to secure their commitment to sustainability and resource stewardship. As one of the first steps to this end, the brewery convened a sixteen-person task force made up of coworkers from various departments. One of the undertakings of this task force was to gather and maintain data on water use, energy conservation, carbon emissions, and other aspects of their footprint. Another undertaking was to establish a list of best practices in brewing that would lessen their ecological impact and to establish a rubric for assessing their success in this effort. The outcome of the task force's work is an annual Sustainability Report. In each report the company establishes goals for reducing water and energy intensity, greenhouse gas emissions, and waste material such as spent grain and sludge. The reports also tell whether or not the company is meeting or exceeding these goals.

As a result of these regular reports, the company's commitment to ecological sustainability became an established norm in the brewery's corporate culture. Once that was established, it was easier for them to take further steps even though the costs would take away from the short term bottom line. In the late 1990s, the employees/coworkers voted to shift to a 100 percent wind-powered electricity program provided by the local public utility. While it is more expensive, the company opted to sacrifice bottom line profits and bonuses for a renewable energy profile. They became the first brewery in the

US to get all of their externally generated electricity from wind. Though most of the wind turbines are located just north of the border in Wyoming, the high volume of electricity flow gave the local Fort Collins utility the inspiration to add their own turbines so they could pitch in. As a result, Fort Collins became the first electric utility in the state of Colorado to offer wind power.

New Belgium is also solar powered. They contracted with Namasté to install a massive array of 810 solar panels on top of the brewery's packaging rooms, which gives the brewery a 444 kwh capacity, or potentially about 17 percent of their electricity use during periods of peak sun. This combination of wind and solar energy made a significant impact in reducing their carbon emissions, which conventionally result from getting electricity from coal powered utilities. Also, throughout the brewery facilities are sun tube skylights that augment electrical lighting with natural lighting. As a nice touch, New Belgium plants fruit trees in their orchard located in the green space surrounding the brewery. Once an employee/coworker becomes vested, the company plants the tree on his or her behalf to help scrub a little more carbon from the air, and the worker gets the fruit.

New Belgium has also innovated ways to convert their waste products into resources. As one would expect, a facility that makes beer out of malted barley and hops requires a lot of cleaning that generates a lot of graywater and waste material. The company sank a considerable investment into converting all that waste water into methane that is burned for more energy at the brewery. As they clean the place, the graywater runoff drains into ponds that were created to be teeming with microorganisms. The bugs consume the nutrients in the waste water and emit methane gas that is captured in a giant space age looking orb located about 300 yards away from the brewery. The gas concentrates in the tank and is then pumped back into a co-generator—a generator that burns gas for electricity as well as for a furnace heater providing warmth. The barley malt sludge is sold to farmers for compost and as fish food. Moreover, the brewery's tasting room and banquet hall are furnished with tables made from discarded bowling alley lanes, its facilities are used by nonprofits and other local organizations for community events, and the carpeting in the offices is made by the famous triple bottom line pioneer, Interface.

For the New Monastics, the company is a treasure trove. The upbeat atmosphere, the enfranchisement of their employees, the celebration of crafts, and their extensive efforts to achieve genuine sustainability, earned New Belgium Brewery its recognition by *Outside Magazine* as the best place to work in America. The people at New Belgium Brewery have created a vibrant sense of place with authenticity that would win the hearts of the Young Americans. Underneath all this vibrancy, creativity, sustainability, and artisanal craftsmanship is a company that makes really good beer. Yet, there is a conventional aspect to the brewery like any other for-profit business. It continues to expand its market share and its product is now sold in all fifty states. Such a wide distribution network from a home base in Fort Collins may prove to be difficult to mainstain in the coming years as water and transportation fuels become increasingly scarce. With all of their millennial efforts to achieve a truly environmentally sustainable profile, only 4 percent of the footprint generated by the beer business is in production. The other 96 percent is in transportation, cold storage, and retail services—all of which are outside New Belgium's control. With its expanding market share and drive for profits, the company is building a new production facility for the East Coast market.

Perhaps in the future the company will take its stock public to generate the capital it would need to replicate the brewery in locations closer to their markets. However, the people at New Belgium are aware of the possible dangers of being taken over by large industrial brewers as has happened in other places. All too often smaller, unconventional or independent companies get assimilated into the corporate hegemony. When we see this happen a feeling of despair sets in, such as many felt when Ben and Jerry's was sold to Unilever, or Burt's Bees was sold to Clorox, or when Portland's famously iconoclastic Voodoo Doughnuts was sold to Yum!, the same company that owns Kentucky Fried Chicken and Taco Bell. We can only hope that New Belgium will not be another casualty of corporate acquisitions.

The New Belgium business paradigm is a hard act to follow. Not many companies are so profitable that they can invest in enormously expensive energy conservation technology and cradle to cradle efficiency. Nonetheless, it is poised to survive when distribution and cold storage costs rise with energy prices and the world goes on a stringent energy diet. If New Belgium, keeps its soul intact and its commitment to efficiency and sustainability, it can survive as it is today—a local business and cultural phenomenon that also makes excellent beer.

The Cooperatives Connection

Local and employee ownership of businesses is arguably the most effective way to re-localize an economy in ways that are fair and democratic. Namasté Solar and New Belgium Brewery in the Colorado Connection are exceptional models of workplace democracy, but not all businesses that claim employee ownership are so democratic or empowering. Most Employee Stock Ownership Plans (ESOPs), for example, are largely very conventional business models that are less likely to have real democratic governance structures. In these models, employees are shareholders whose shares are placed under the control of a third party trust. Management personnel of the ESOPs are answerable to their board of directors and the board is accountable to the trust, not the employees. In this way, ESOPs look more like 401(k) plans, and more direct grassroots economic enfranchisement of people in the local community are cooperatives.

Currently about 750–800 million people in the world are members of cooperatives. In the US, there are about 48,000 cooperatives with over 120 million members. Traditionally, cooperatives were created to pool resources from within a community to provide goods and services that are necessary for the wellbeing of the people. Modern cooperatives date back to the populist movements of the mid-19th century and the heyday of monopoly capitalism. Poverty, low wages, low prices for farmers and dismal working conditions compelled people in communities to pool their resources and form cooperatives as alternative business models. As with organized labor, communities that were organizing cooperatives had discovered that collective action and pooled resources translated into protection against the otherwise grim, Darwinistic struggle for survival in the market place.

The most common forms of cooperatives are consumer cooperatives and producer cooperatives. Consumer co-ops are owned by the people who buy the goods and services provided by the co-op, and producer co-ops are owned by people who work for them. There are far more consumer co-ops than producers. This is due to the fact that consumers constitute a wider base within the community and therefore offer more resources for capitalization. Traditionally, most producer cooperatives are owned by cooperative associations in which independent producers—farmers, artisans, or people engaged in other small scale trades—form a collective to combine resources and purchasing power in buying raw materials or energy. Producer cooperatives could also be worker cooperatives in which membership and ownership is held by employees of the company, but these are less common in the US economy.

Though cooperatives remain a very small piece of our economic system, they hold much promise for a future re-localized economy that is coping with energy descent. Worker cooperatives are less prevalent than consumer co-ops. They are promising because they are not created specifically for growth and profit maximization, but to serve the community. Even more promising are networks of cooperatives that are working together to help each other develop and survive in an otherwise troubled and threatened economic environment.

One such network, known as the Evergreen Cooperatives, can be found in Cleveland, Ohio. Currently the Evergreen Cooperatives consist of the Evergreen Cooperative Laundry, Ohio Cooperative Solar, and the Green City Growers Cooperative. This network is a model of how community leaders set out to economically rebuild part of the city that is troubled with high unemployment and low income and just happens to also be located in the heart of Cleveland's health care industry. The health clinics, University Hospitals, and the Veterans Administration's Medical Center comprise what the Evergreen founders call "anchor institutions" that employ over 50,000 people and constitute the dominant economic force in Northeast Ohio. The Evergreen model is a unique structure built around these anchor institutions that support new businesses specifically created to generate decent livelihoods in an otherwise depressed local community.

In 2005, The Cleveland Foundation, a local community foundation, began an effort to help deal with the poverty and unemployment, and to rebuild a vibrant inner city community. The Foundation formed partnerships with local community-based organizations, community leaders and the anchor institutions to work on various development projects. The flagship project was the Evergreen Cooperative initiative that was launched in 2007. The initiative was systematically created to be locally self-sufficient and draws on some precedents of employee-owned businesses in Ohio's steel industry. The one-time startup financing from loans and grants was aggregated to provide capitalization, but from there the co-ops run on their own revenues. The mission is to focus the procurement needs of the anchor institutions to locally sourced and owned companies. The initiative forged a mutual support structure that would spread out the concentrated wealth of the anchor institutions in exchange for goods and services provided by a group of community-based cooperatives. This is an example of a community successfully revitalizing its economy, albeit though not by the conventional means of imploring large companies to locate their businesses in the area. Instead the people of Cleveland built new institutions from the ground up, not only create jobs, but to more equitably redistribute ownership and wealth in a mutually beneficial symbiotic way. Evergreen creates jobs and opportunities for ownership, then trains people from the local community to fill the positions and channels multi-billions of dollars spent on business services by the local health care industry directly into the communities and households that need it.

In 2009, Evergreen Cooperative Laundry (ECL) and Ohio Cooperative Solar (OCS) were the first two companies created as part of the initiative. ECL specializes in providing environmentally clean laundry services for the local hospitals and health clinics. OCS specializes in clean energy and weatherization projects for both the anchor institutions as well as for neighborhood residents. A year later Green City Growers (GCG) was created to produce food for the community in a 10 acre hydroponic facility of which 5 acres are under glass. GCG is the largest single urban food production facility in the country and can produce over 3 million heads of lettuce every year. Together these cooperatives created over 100 new jobs, most of which are also enfranchised with a shared ownership of the company—ownership by those actually doing the work. And built into the company charters is institutional language that specifies that the companies cannot be sold.

The cooperatives remain going concerns that could gain momentum. As a network of enterprises, it could evolve into a full-fledged local commonwealth. The people of Cleveland are hopeful that these cooperatives are just the beginning of a much large spectrum of locally based and locally owned enterprises that are chartered specifically to create local economic stability and ecological sustainability. It is still in its early stages of development, but has great potential to evolve from a few boutique enterprises to scale out into all aspects that are necessary for a complete commonwealth with over $3 billion in procurements by anchor institutions as its foundation. The Evergreen model is a successful alternative to other attempts at inner city revival in which low cost loans to small businesses and job training programs did not generate results. The initiative has also created a fund in which 10 percent of the companies' profits are allocated for other community regeneration projects.

A discussion of cooperatives would not be complete without mentioning the Mondragón system in the Basque region of Northern Spain. In many ways, the Mondragón Corporation is ground zero for the networked cooperative movement. The founders of the Evergreen system went to Spain to study the business models there and were inspired to recreate something similar in Ohio.

Mondragón evolved from a small project created in the early 1940s by Father José Arizmendiarrieta (Arizmendi for short), a Catholic priest, humanist, and a believer in science, technology, and education. As a newly ordained priest assigned to the small town of Mondragón, Arizmendi was dismayed by the poverty and unemployment he saw there. Much of it was caused by the long and tragic civil war that did much irreparable damage to the people and culture of Spain. Looking to be proactive, Arizmendi saw community-based economic development as the solution to his town's problems. To that end he created a humble education center and a restoration of what was a local Basque tradition of consumer and worker cooperative enterprises. Throughout the 1940s, Arizmendi's school successfully educated managers, engineers, and skilled workers. And by the late 1950s, they created the Mondragón Worker Cooperative Federation that linked a number of local, employee owned businesses together into a network. Later they added a credit union to provide financing for more development.

The story of Mondragón, however, is more of a cautionary tale of how business models can be thrown off their rails by their own success. Today the Mondragón Corporation is a €15 billion ($20 billion US) enterprise that employs about 84,000 people. It now has a global profile with production facilities in low wage countries in order to increase its international competitiveness and market share. Once a model for local, human centered economic development, Mondragón is firmly embedded in a global, neoliberal capitalist system. The Evergreen founders were impressed with Mondragón's success, but the rubric for measuring this success—growth, profit, market expansion—is the same as any conventional capitalist enterprise, aside from its inspiring history, is of little use for the New Monastics. There is nothing like the lure of financial wealth to cause businesses to lose sight of their mission and compromise their values. It is not at all helpful to deceive ourselves or to pretend to being doing one thing while in reality doing another.

Employee ownership models like Namasté Solar, New Belgium Brewery, and the Evergreen Cooperatives can also be seen as a viable alternative to labor unions that are wavering and under constant attack. Around 7 percent of the American private sector labor force remains unionized and as the industrial base of our economic system decays along with energy descent, workers can become empowered in new industries through ownership rather than collective bargaining agreements. The models for the economic empowerment of local communities are valuable treasures for our post carbon commonwealth. Part of the ongoing work that needs to be done to create this commonwealth is to create many more worker cooperatives as they still constitute a small percentage of the overall cooperative economy. Worker ownership anchors business enterprise to the local community and as such provides badly needed stability, which will become increasingly necessary in turbulent times ahead as energy descent accelerates.

The Health Care Connection

Access to affordable and quality health care remains a troubling and growing problem for millions of Americans. There are a few ways to approach this. One is the more traditional way of building a national, public sector, single payer system for health insurance. This was suggested as part of the national health care plan, but was crushed by health industry lobbyists in Washington. If we continue to fight for this—as it merits—at the same time we could be creating local alternatives that would empower us by eliminating our dependency on insurance companies for our health care. Once we achieve independence from large and powerful insurance companies, we will be in a stronger position to demand reform legislation at the federal level.

However, there is a treasure buried in the otherwise mangled Patient Protection and Affordable Care Act of 2010. The bill contains some provisions that could make coverage more accessible and that authorize individual states to create their own nonprofit health plans called Consumer Operated and Oriented Plans (CO-OPs). In this model, providers are governed by the customers who are paying the insurance premiums. As the CO-OPs are consumer centered, they are created specifically to give premium payers more control over their health care coverage and better plan accountability. The federal legislation also authorized the federal government to create a $3.8 billion fund as a loan program to help qualified CO-OPs get started. To oversee the program, the legislation has also created a fifteen-member Program Advisory Board to oversee that the CO-OPs are consumer operated and controlled.

To be eligible for the federally supported CO-OP program, the nonprofits created in the states must be licensed insurers by the state. To assure their independence the CO-OPs cannot be owned or controlled by, or receive more than 25 percent of their funding from, an existing insurance company. They can receive grant money from state or local government agencies, but cannot be controlled by these agencies, nor receive more than 40 percent of their funding from them. As soon as the program was opened, applications for starting these CO-OPs with federal loans poured in from all around the country. The laws on insurance coverage varies from state to state and there are some legal tangles about whether these entities must be strictly consumer owned, whether they must follow strictly the guidelines for cooperatives, or function as nonprofits. There is nothing, though, that would prevent these insurance companies from being strictly private, nonprofit entities that are governed by a board structure comprised of customers and dedicated to the cause of providing excellent, low cost coverage. Such CO-OPs are sprouting up around the country and have the potential to create significant changes in the health care industry.

As we strive for good health care plans and legislation, it is important to keep in mind that many of the traditional models no longer apply and we'll have to become more creative in our efforts. Our educational institutions are cranking out degrees like never before, and their graduates are filling the ranks of the unemployed and uninsured. For many of those with jobs, their careers do not follow the traditional model of full time employment supported with health care benefits and pensions. Instead an increasingly large percentage of our labor force is made up of people who are trying to patch together a livelihood by working multiple part time jobs, temping, or working as independent contractors. Businesses are less stationary and this is causing our labor force to become more independent or itinerate as they shift from one project to the next. None of these shifting employment circumstances is likely to provide the workers with anything but the don't-get-sick health insurance plans.

The Freelancers Union is attempting to fill this gap for independent workers on a number of fronts including health care. The organization is a nonprofit that was created in 2001 to help independent workers through political advocacy and support programs. Its most important support program is to provide health insurance for its members, which as of this writing stands close to 170,000 though most of them are in New York. Initially the union helped its members obtain insurance coverage as a nonprofit insurance broker based on the idea that with power in numbers they could more effectively negotiate better premiums and coverage for its members than they could get individually.

In 2008, the Freelancers Union started its own for-profit insurance company, Freelancers Insurance Company, to provide state-authorized insurance directly to members rather than brokering the services from other companies such as Blue Cross Blue Shield, though only in the state of New York. The union's company asserts that its premiums are more affordable than those of the big insurance companies as the company exists to serve its members not shareholders, which is exclusively the Freelancers Union nonprofit. Some of the members felt that there was an abrupt and confusing transition from having their insurance brokered by the union's to the union becoming their exclusive insurance provider. Although the transition seemed quite bumpy, the union continues to offer group rate insurance that is more affordable than that available from their giant corporate competitors, and the ranks of its membership continue to grow. Freelancers Insurance Company insures about 25,000 independent workers in New York.

The citizens of Vermont have taken yet another approach. In 2011 they passed a new law that authorized the state to form a state-run, single-payer health care system. The critics of this legislation argued repeatedly that getting a public option would never be possible as it appeared to be un-American. Nonetheless the activist group that pushed for the legislation, the Vermont Workers Center, pressed onward and the bill, *Act 48: Relating to a Universal Unified Health System*, was signed into law. The central principle of the reform bill is that access to health care is a human right on a par with access to education and protections against racial or gender discrimination. This principle and several other supporting principles were integrated into the statutory language of the legislation:

> It is the intent of the general assembly [of Vermont] to create Green Mountain Care to contain costs and provide, as a public good, comprehensive affordable, high quality, publicly financed health care coverage for all Vermont residents in a seamless manner regardless of income, assets, health status, or availability of other health coverage.

Green Mountain Care is a state run plan with several programs created to implement the provisions contained in Act 48. The programs provide comprehensive health care coverage through the state and its private sector partners. The programs are geared primarily for those who are uninsured and are unable to get insurance coverage because they are unemployed or due to low income. One such program is the Vermont Health Access Plan (VHAP). To be eligible for this program an individual has to be in a very low income situation with a maximum household income of $43,000 or less for a family of four and have been without insurance for twelve months or more. In this program, a person is not immediately eligible upon loss of job and insurance coverage, or loss of coverage due to divorce.

Those who are not eligible for VHAP for some reason but have nonetheless been uninsured for twelve months, can enroll in another plan, Catamount Health. This program has insurance plans that are either provided by the state or by private sector insurance companies. For both VHAP and Catamount, coverage includes regular checkups, hospitalization, screenings, emergency care, chronic disease care, doctor visits, prescription medicine, and other services. The premiums are on a sliding scale based on income levels and household size that range from $60 up to $513 per month.

Vermont's Green Mountain model also has programs to protect uninsured children and pregnant young women. The model is not perfect and it also has the fingerprints of the private health insurance industry all over it. For one thing, the VHAP program was clearly created for those who would not be able to get private insurance due to unemployment and no income and therefore would not even be in the market for insurance. As it is targeted for the already low income and uninsured population, the Green Mountain model does not pose any competition to the private insurance industry. It does have a program for uninsured middle-income families, but these are very expensive full pay premiums and are provided by private companies like, once again, Blue Cross Blue Shield.

Obviously health care reform has a long way to go in the US. The Freelancer's Union and Green Mountain are taking the initiative to make health insurance available and affordable to those who have fallen through the cracks of the conventional system. Yet the programs overall are still under the control of private sector insurance giants like Blue Cross Blue Shield. The inroads they have made, however, are feet in the doors that can open up to wider possibilities. The fact that this legislation even exists means that state laws can be amended and amended again until big corporate insurance companies are pushed out entirely. For this to happen, though, would require a very active, educated, organized, and vocal citizenry.

Access to health care is heavily impeded by corporate influence on legislation. Both at the federal and state levels reform legislation is troubled by the omnipresence of corporate money. This is even more reason why an active citizenry should be pushing up against these structures of moneyed power from the community level as the Home Rule initiatives supported CLEDF and others. And if people find themselves overwhelmed and rendered ineffectual by the magnitude of the efforts that will need to be taken, there are still things people can do to change the most ancient and primordial of all economic institutions—their households.

The Household Connection

The word "economics" originates with the classical Greek word *oeconomia* meaning "householding." In ancient Greece, agriculture constituted the bulk of production and work was primarily carried out by the small *oecos*, meaning the household estate or family farm. The Greeks were independent-minded and held a well-run, self-sufficient family estate in high regard. The proper management of the household estate was so important to the Greeks that the famous general, teacher and historian Xenophon (434–355 BC) wrote a treatise on the subject titled, *Oekonomikos*, arguably the first text on economics in the history of Western civilization.

The definition of what constitutes a household as an economic institution has evolved since Xenophon's day and as we prepare for a world without oil it will continue evolving. Energy descent will force households to undergo a transformation just like every other economic institution. There are a tremendous number of ways in which people are actively changing how they eat, live, travel, work, care for each other, and spend their free time as they are compelled by the coming scarcities of energy. A central theme of this book has been that the end of the Oil Age is coming and the fact that we will have to reduce our energy and material consumption is a certainty. Consumerism remains a dominant force in our culture and as long as we allow it to be so, then we are headed for intractable problems and conflict. Those who refuse to acknowledge the coming necessity of lifestyle changes away from high levels of consumption are likely to view scarcity as something that will require sacrifices that someone else will have to make, not themselves. This attitude will most certainly bring about class conflict as wealth and income distribution in America is more polarized than any time in modern history. It is hard to imagine anything good coming out of such conflict, even for those who are the most affluent, as they will be forever trapped in a continuous state of fear and will be forced to watch their once peaceful communities go up in flames.

A more sane and just approach could be to continue exploring ways in which scaling down energy and material consumption does not mean tossing entire segments of our population into poverty and suffering. This approach must also be accompanied by a reevaluation of what in the popular imagination is considered to be the good life. For us in the US, embracing such a concept would require that we redefine what we mean by economic value, prosperity, as well as redefining what are the most important needs that have to be met. But as I have also argued, we cannot "lifestyle" our way out of environmental and social problems. It is true that we can choose to live in smaller more energy efficient homes, opt for different modes of transportation, and pursue less materialistic ways of living, but if we did these things in significant numbers, conventional business earnings would crash and along with that would go the stock markets, government tax revenues, jobs, and a general acceleration of economic crisis conditions. In other words, as always, these are broader institutional problems.

Institutional problems require institutional change as their solution. Redefining household institutions in accordance with a new conception of the good life characterized by lighter and simpler ways of living could only work if business institutions were also redefined away from endless growth and profit maximization, and only if we were to let go of our own expectations that financial institutions will always pay off with higher returns. In other words, we have to be evolving all these institutions at the same time and in the same direction.

For many people, working toward such a great transformation is a welcome change. More and more people are choosing to simplify their households by building smaller homes that are less expensive, cheaper to maintain, more energy efficient, and consistent with urban density plans to locate people closer to amenities. As they spend less time in the rat race of producing and consuming more and more, they can spend more time doing different kinds of work: urban food production and crafts, or spending time retraining themselves to become more self-sufficient, relearning how to make repairs, how to do their own home improvement projects and restorations, or spending time bicycling rather than in crowded traffic jams seething with road rage and suffocating with carbon effluents.

As the carbon economy winds down, so will the money economy. This is not to say that we will eliminate money, but rather that we can reverse our relationship with it and reenvision the role it plays in our lives. Most of us are ruled by the money economy. We feel trapped because we need to make more and more in order to keep from falling further and further behind everyone else. Under these intense conditions of consumerist competition or competitive acquisition, too many people feel a sense of alienation and shame when they lack the consumer goods that others seem to be enjoying. But the overall energy/economic descent will change all of that anyway, so it seems logical that we will be creative in coming up with alternative simpler lifestyles that will eventually become the rule rather than the exception. And as we become more self-sufficient with our time, we break free of the grip that money has had on us and money will be eventually restored to its original purpose of facilitating exchange and nothing more. That is, money can remain as a tool for efficient exchange rather than an institution that dominates our lives.

As the primordial economic institution, the household will remain as the nexus of all our economic activity. The physical, mental, and spiritual well-being of each member in it is what our economic activity should be geared toward and what should also be the paramount guiding principle of our commonwealth. This means a conscious and mindful effort to redefine ourselves.

Conclusion

Recall that the scientists at the Swiss Federal Institute of Technology estimate that a sustainable daily amount of energy consumption would be the equivalent of about 2,000 watts per day per person.[57] That number will continually decline as we add people to the planet. But even at the current level, that amount would require that Americans would have to reduce their energy consumption by 75 percent. Without significant institutional transformation that would be catastrophic. Putting this another way, if in our holistic view of things we see our economic system as like an organism and that organism is suddenly required to lose 75 percent of its weight, it would most surely die. If we passively stand by and wait for our economic system to collapse because it simply ran out of gas, then the future for most of us will be very bleak. The organic whole needs to evolve into a new genus and new species of a body economic. Our primary task as the New Monastics must be to evolve our economic system into a new and very different commonwealth through a process of local institution transformations.

These transformations begin with an informed vision of place. This place is economically stable because it has firmly set its roots in the local community through local ownership and control as is done with New Belgium, Namasté Solar, and the Evergreen Cooperatives. It is ecologically stable and permanent because the institutions that organize the local economic activity are governed by better rules and shared strategies for species protection, land and water conservation, clean energy, sustainable carbon emissions, and other forms of stewardship. The key that sets these New Monastics apart from business and government as usual is that they have concretized their rules of governance into their charters, statutes and shared strategies. Like meta-economists they are crafted with the grammar and syntax of ecological permanence, the joys of good work, stability, and genuine sense of fairness and justice.

There has yet to be invented a clean litmus test that can be used to determine which models are genuinely useful for a steady state commonwealth versus those models that are just more business as usual hiding under green wigs. It seems that the more we look for this, the more elusive it becomes. As I took to the road and interviewed these people about their work, I discovered that many of those whom I expected to be authentically different and forward thinking turned out to be shams, while those I stumbled upon serendipitously turned out to be the most compelling, and many of those turned out to be for-profit entrepreneurs whose business models on the surface might seem indistinguishable from mainstream business. Oddballs, as Morris Berman says in *The Twilight of American Culture*, but not necessarily renegades. Rather I discovered individuals with a deep commitment to certain practices that resonate with the meta-economist visionaries.

There is no blueprint for a shadow economic system as described by David Ehrenfeld, nor is there any one particular institution that constitutes the "it" model. What we find instead are unique gems and treasures worth preserving or pieces of a jigsaw puzzle, the image of which is yet to be determined and perhaps unknowable until it is defined by future historians. What makes them worth preserving is that they point in the direction of a steady state and stable commonwealth. Each piece alone is practically meaningless, but taken together they are like a piece of genetic information that will comprise the DNA of an organism that has yet to evolve into being. The New Monastics described here are a very short list of multitudes of initiatives taking place everywhere whose emergent properties will give rise to a new system—a new commonwealth. But the work for each of us individually lies in crafting the grammar and syntax that is written into the governing structures of all economic institutions through their Articles of Incorporation, Partnership Agreements, Bylaws, Mission Statements, or simply informal rules. In time, these rules become naturalized as cultural norms such that, as with a familiar painting hanging on an office wall, people do not even notice their existence. As such, these become an integral part of our living culture.

CHAPTER SIX

EDUCATION AND OUR GREAT TRANSFORMATION

One of the hallmarks of economic globalization has been an expansion of the open market to every corner of the world. Relying on market forces to allocate oil and other resources among the world's population is dangerous as it will only worsen the already troubling conditions of economic inequality. The more affluent members of the world community will have an easier time not only paying for those more expensive resources, but also paying for the necessary infrastructure for alternative energy sources. What may be considered progress or prosperity in one location, no matter how sustainably or equitably it was achieved, could nonetheless create problems for others elsewhere. The schism between classes will grow wider as the poor are rationed out of the market for vital resources for no other reason than the lack of their ability to pay for them. This most certainly will lead to conflict. We as educators have a responsibility to emphasize humanistic education, and compassion, and to foster a greater global awareness of the impact resource scarcity will have on us all. With a more global consciousness citizens around the world can work toward building international institutions that can help us deal with energy scarcity in ways that are politically stable and just.

Developing a global consciousness is not an endorsement of economic globalization, nor does economic re-localization imply reverting backward into parochialism, nationalism, or isolationism. All too often the way citizenship and civic responsibility are taught in the US is not to foster social creativity and problem solving, but rather to instill a sense of ethnocentrism and patriotism. This dulls people's sense of awareness and concern for others and, again, this is dangerous. Instead our educational institutions need to be crafting an appreciation of global citizenship in our students as well as working toward revitalizing our local communities. As Dewey and Geddes taught us, public spiritedness is both a process of individual development and civic regeneration. As the individual person grows intellectually and spiritually, the sphere of that person's experiences will expand beyond community boundaries, beyond national boundaries to a global scale. The essence of humanitarian education is to foster a sense of the goodness that comes from contributing to one's own community to further the cause of social justice and stability. And as citizens develop this sense and regenerate their communities, the world becomes more stable and just. A more stable and just world will then reverberate back to each community, reinforcing these values. As we become active citizens in our communities, we also become active citizens of the world.

To develop a sense of global responsibility while the world slides into energy descent will be an important challenge. It will require not only awareness, but insight and wisdom. As educators, we have to decide what kind of wisdom we want to cultivate in our students. Nel Noddings, in *Educating Citizens for Global Awareness* (2005), asks, "Dare we ask our students to consider adapting economic moderation as a virtue?"[58] It seems that when she asked this question of her students at Stanford University, the response was affirmative. Though Stanford students may not be a representative sample of the US population overall, it nonetheless begs the follow-up question of how to go about pursuing such moderation. In my own classes, I've had my students do similar thought experiments and ask: How do we live peacefully and equitably in a world with 75 percent fewer resources than what we have now? But in this experiment I also set the parameter that technological solutions are off the table. The idea behind this thought experiment is to get students to think beyond the knee jerk cliché that technology is always the answer and will always be there to save us. Noddings continues, "The Charter of the United Nations refers to 'fundamental human rights,' to the 'dignity and worth of the human person,' and to the 'equal rights of men and women and of nations large and small.' Its Universal Declaration of Human Rights was adopted in 1948 and should be studied by all aspiring global citizens . . . The world has fallen pitifully short of upholding these ideals."[59]

Just because an ideal or aspiration seems beyond reach at the moment, does that mean it should therefore be abandoned? Perhaps it would seem so, if only to the most cynical among us. But we should exercise caution in deciding for others what is in their best interest. Teaching global citizenship and global responsibility cannot be boiled down to one particular ideology, but it can strive to achieve unity within diversity and find a set of common themes that rise to the surface. And then perhaps those themes can begin taking on the form of a global consensus that can be shaped into institutional statements that have specific meaning and relevance for every community. To be successful, however, this effort will require genuine and thoroughgoing educational reforms, particularly in public schools.

The original idea behind public education was to cultivate literacy and a sense of civic responsibility in order to achieve a functional democracy. This has changed. Public education is more and more being seen in the same way as other public services that exist to serve the needs of those who lack means, such as Medicaid, public housing, public transportation. Disparities in income levels are also causing chronic budget shortfalls and failing support for public education. The more affluent among us feel that they have already spent their money educating their own children in private schools and feel no responsibility to pay taxes for educating the children of others.

As these conditions deteriorate public education gets swallowed up by corporate interests. Schools increasingly are forced to turn to corporate patrons for funding, which in turn demand that the curricula focus on specific outcomes that are generally of interest to these companies. Administrators are often seen tagging along with corporate CEOs, announcing with great fanfare their public-private collaboration. And as we increasingly rely on corporations for funding students naturally are led to understand that the education process amounts to little more than workforce training. What fades away is the emphasis on civic participation—the process of creating an educated, critically thinking citizenry armed with the power and will to build new alternative institutions. This is happening at a time when civic participation and global awareness are needed more than ever.

This must change if we are to live peacefully in a world without cheap energy. Civic participation, civic regeneration, and institution building are going to require people of all ages to have at least a basic understanding of why these are important and what role they can play in the regenerative process of change. Educators need to have their students working on social change models. Such models need to be informed with wisdom, not self-interest or conventional pieties and developed with a deeper sense of moral and civic virtue. This entails developing programs that give students a sense of working for the common good. In addition to problem solving using social change models, education needs to be directed at cultivating sensitivity to the well-being of others everywhere, not just in their own community or country. They have to be trained to organize others around economic and ecological issues that energy descent will bring. Some of that work will be in public service, some of it in consensus building and action plans, and some of it will be doing artistic work and cultural productions that also serve the common good. And to be successful community citizens and citizens of the world in the post Oil Age period, we are all going to need new skills.

Developing Our Skills—New and Old

One of the most remarkable attributes of our species is that we can adapt to nearly any condition we find ourselves in. But adaptation is not merely learning to live with what we've been given or accepting our circumstances, it is the ability to transform ourselves and evolve. Our great transformation is dependent on our ability to be individually, socially, and culturally malleable as the conditions of our world change.

Transforming our economic system into something new and better will require the development of educational programs aligned with this effort. Moving toward a livable future will require that people develop new skills and technologies, create new cultures, foster creativity, and raise awareness. Educational institutions need to develop new curricula and workshops that will provide tools and guidelines for people to relearn how to become functioning citizens and members of their communities because economic re-localization will require it. People of all ages will have to develop entirely new capabilities to design, heal, manufacture, grow, repair, write laws, develop curricula, plan, rebuild; to become board members, visionaries, scholars, artisans, intellectuals, shared strategists, gardeners, woodworkers, metalsmiths, canners, seed librarians, as well as to become the visionaries of their own generations.

The skills themselves, as they are developed, will become the supporting substance of a new commonwealth and will evolve into a new culture. With the evolution of each community, the global community will evolve as well. In her essay, titled "Differing Concepts of Citzenship: Schools and Communities as Sites of Civic Development," Gloria Ladson-Billings writes:

> I am convinced that until students begin to see democracy work in their own local communities, their ability to work for it as a part of the common good and worthwhile global strategy is unlikely to

materialize. One of the best examples of this strategy is exemplified in the work and legacy of Myles Horton, founder of the Highlander Folk School, which began as a center of training for union organizers. Horton understood that the people with the problems are also the people with the solutions . . . labor organizers, civil rights activists, antipoverty workers, and other community workers assembled to solve local (and ultimately national) problems. . . . Some of Highlander's "graduates" included Eleanor Roosevelt, . . . Martin Luther King Jr. . . . Rosa Parks . . . They understand the way our local problems and solutions can have a global impact.[60]

Those who go on to build new educational institutions and programs in the spirit of the Highlander Folk School will be among the New Monastics. There should be less emphasis on training young people for traditional careers. Changing economic circumstances are rendering traditional education-to-career tracks obsolete.

More emphasis needs to be placed on relearning the lost arts of practical life skills, as well as crafting new ones that will be treasured by neighborhoods and communities. Though the grim reality of deteriorating public support for education is a challenge, educational institutions and communities working together with a clear vision can have amazing results. As people learn to do these things for the common good, the specific institutional forms will naturally sprout and grow and network with each other. As they do a new commonwealth will naturally evolve. What it evolves into, however, will depend on the efforts for change we pursue today.

Reforming Economics Education for the 21st Century

The litany of indictments of failing education systems is familiar: high dropout rates, poor literacy, innumeracy, and underdeveloped critical thinking skills. But there are other less mentioned ways our education systems have become dysfunctional. Educators ought to be those who foster the development of new generations of fresh thinkers, creative souls, and new problem solvers. Instead, however, our educational institutions do more to indoctrinate our young with stale conventions and outmoded paradigms. And the discipline of economics is among the worst, with its moldy clichés and slavish adherence to neoliberal ideology. This is tragic given the potential for positive change if citizens were empowered with real knowledge about how economic systems function, or how they fail to serve the basic needs of people. But real world understanding is nearly impossible to find for those trying to make their way through the thorny hedge of academic economics.

If we were to boil down mainstream economics as it is taught to college students, we would get basically two things: microeconomics and macroeconomics. Most of the specialty courses are extensions of one or the other or both of these. At the core of microeconomics is an assumption that the key elements behind the forces of supply and demand in the marketplace are choices. It exists in a make believe world in which institutions or structures of power are almost completely absent. The marketplace is idealized as an open playing field in which individuals make choices based on a principle of individual satisfaction in which everyone is believed to be driven by a single impulse—the pleasure principle. We make choices as to what to buy or consume by doing a kind of cost/benefit analysis in which we weigh the benefits or pleasures derived from incrementally consuming more of some good against the direct cost (negative pleasure) of obtaining it, or against an direct cost of whatever other thing you had to forgo to obtain it. In microeconomics, everything is about being free to choose—to choose among competing products or competing career paths—in such a way as to yield the most satisfaction.

Companies are said to engage in the same cost/benefit choice-making behavior as they choose among products and resources that will generate the most profits. Though a company's goal is different—profit maximization rather than personal satisfaction—the logic is exactly the same. The decision of what or how to produce is made by weighing the costs, direct or indirect, of producing incrementally more quantities of some good against the gains from additional sales. The decision on how much to produce is based entirely on the profit maximization outcome of this cost/benefit analysis.

Though to some degree people and companies do this, defining choice-making behavior in this way fails to take account of the institutional forces that set the parameters of those choices. Institutional forces are either completely ignored or only addressed tangentially in textbook economics. Individual choice-making, in other words, is seen as natural and institutional forces as unnatural and unwelcome intrusions. To the extent that institutions exist at all in these models, they are largely seen as outside forces that tend to gum up the workings of an otherwise perfect universe of free individuals. The only real exception to this is the limited role of government "intervention" that can be tolerated only if it can be proven to promote more economic growth. It stands to reason that if you can purge any understanding of the power of social institutions, you can also eliminate a person's desire to change or replace those institutions. In this way, textbook economics serves to uphold established conventions and pieties: indoctrination. Critical thought is pushed aside to make room for apocryphal stories of how human selfishness in an unfettered market environment leads to social progress.

In November 2011, a group of Harvard students walked out of an introductory course in economics taught by Gregory Mankiw, formerly advisor to President George W. Bush and Mitt Romney. The students' walkout was in part to show solidarity with the Occupy movement, but also because of the heavy bias toward neoliberalism that is embedded in standard economic theory. One of the protestors complained that the course was very indoctrinating and did not encourage diversity of views beyond the core assumptions that support neoliberalism.

These core assumptions of consumer and business behavior are taught not just at Harvard, but in virtually every introductory course in economics in the country. They are held to be the substance behind the forces of supply and demand in the marketplace. As students of economics progress from introductory courses to intermediate and graduate levels, the same core assumptions about individual choice-making behavior remain intact, though the modeling techniques become more mathematically rigorous. And at the graduate level, the level of rigor is ridiculous. Students are expected to apply advanced calculus, linear algebra, and multivariate regression analysis to model things like how people and businesses choose between hot dogs or pizza. When the choice involves more than two options, the mathematical complexity goes beyond madness. For example, take this quote from a standard textbook in advanced economics:

> When there appear in an objective function N > 2 choice variables, it is no longer possible to graph the function, and though we can still speak of a hypersurface in an (N + 1)-dimensional system. On such a (nongraphable) hypersurface, there again may exist (N + 1)-dimensional analogs of peaks of domes and bottoms of bowls.[61]

This passage is set amidst pages and pages of sets of linear equations and Hessian matrices that are somehow supposed to shed light on how we make choices between hot dogs, pizza, plus some third thing that makes the function ungraphable. Though the mathematics can be understood, its relevance to how people and businesses behave remains an opaque mystery. Macroeconomics is something of an extension of microeconomics. Though macro does make some attempt at connecting to something that actually exists in the world, it does so with the same level of mathematical absurdity. However, all the concepts in macro are centered on a single premise: the prime directive of all economic systems is to maintain endless economic growth.

At some point students may begin to question why economics is taught using such alienating, irrelevant, and bizarre methods. The answer they are typically given is an unsatisfactory tautology: "We teach it this way because this is the way it is taught." In other words, it has no meaning or relevance whatsoever other than "This is the way academic economists build their careers."

However, some of the rare critics in the business have offered more precise explanations. Philip Mirowski, in his book *Against Mechanism* (1988), argues that economists do this because they suffer from a condition that drives them to emulate the level of mathematical sophistication of physicists. Mirowski call the condition "physics envy"[62] and academic economics has evolved into nothing more than endless variations on pointless models in order to achieve the status of rigorous science. Yet even with all their splendid mathematical rigor, on the subject of global economic and ecological conditions that are becoming increasingly unstable, academic economics remains appallingly mute.

Instead of giving helpful concepts and guidelines for how to rebuild a commonwealth, economic discourse is loaded with stale and dreary arguments over meaningless froth. Or worse, it remains trapped in endless debates over mathematical formalisms that mean nothing to anyone outside the club. The key to staying in the club is to use increasingly inaccessible mathematical techniques such that it intimidates anyone but those who are most devoted. That is, economists use arcane mathematics to intimidate their critics and keep them outside the margins. Very few would want to bother spending years developing their math skills and analytical abilities only to be able to state a fact which is obvious to everyone else anyway—that textbook economics amounts to nothing useful whatsoever. Thus only the club members are the devotees and the hopelessly quixotic heterodox economists who remain grumbling and growling from the sidelines. As I argued in *Mindful Economics* (2008), if educating people about the economy is not the purpose of academic economics, what is? According to notable scholar and critic, E. K. Hunt:

Economists did succeed in erecting an impressive intellectual defense of the classical liberal policy of laissez faire [free market capitalism]. They did it by creating a giant chasm between economic theory and economic reality, however. From the 1870s until today, many economists in the neoclassical tradition have abandoned any real concerns with existing economic institutions and problems. Instead, many of them have retired to the rarefied stratosphere of mathematical model building, constructing endless variations on esoteric trivia.[63]

This is unsettling considering how important the real study of economics actually is. The importance of it lies with the fact that economic forces determine how we live. In every election cycle what tops the list of issues that are of concern to citizens are economic issues. Labor markets conditions determine the circumstances of our employment, and those conditions determine the affordability of food, housing, health care, and all the other necessities of life. Students should be pushing for more accountability and reform in the way economics is taught. But the agitation should be coming not only from the students. Accreditation boards, professional associations and funding organizations should be asking these questions as well.

Our efforts to build a new commonwealth at the end of the Oil Age need wisdom, insight, compassion, as well as sound scientific reasoning. True educators who encourage independent thought and inquiry will themselves become the New Monastics who will carry these as treasures for future generations who will need them to survive in a world without oil.

CHANGING THE WAY WE APPROACH CHANGE

The problem with describing a post carbon economy is that such a thing does not yet exist. Uncertainty abounds as the once thriving global economy is literally running out of gas and none of us can be sure what will come next. We can, however, be certain about one thing: the primary measures of success—economic growth and monetary accumulation—will sooner or later become obsolete. When that happens, everything will change. As we shift away from perpetual growth, we will find ourselves trying to stay afloat in uncharted waters. The way we produce and consume things, our values, our aspirations, our government policies, and pretty much everything else will undergo a transformation. The transition is not going to be easy, and this is probably why so many of us believe that we need to do things differently—only to fall back on variations of business as usual. Many others still will mock true visions for change as idealistic dreaming and cling to the cynical notion that people will never change even if the world compels them to.

This notion is neither helpful nor true. Everything in the world around us is in a state of impermanence and transformation. Even during the shortest imaginable stretches of time nothing remains absolutely constant except perhaps our comforting illusion that it does. Meanwhile we are pulled deeper into a vortex of social crises as the limitations of our physical world render business as usual and peace an impossible combination. Yet our societies at large keep pretending that it is not an impossible combination even when evidence to the contrary is staring everyone straight in the face. As the old saying goes, insanity is doing the same thing over and over again but expecting different results.[64] Flux, transformation, and impermanence are real and immutable, but what is also certain is that if we want to survive, we need to change how we approach change.

In light of that, I want to urge caution. We need to be cautious about throwing ourselves into the next big movement that is going to shake the world, or behind the next charismatic political leader who comes along with promises to save us all. It is never that simple. Old systems do not suddenly vanish nor do new systems suddenly burst into the world like a volcanic eruption. Rather, the populations of the world are experiencing an evolutionary transformation in which the current system is becoming increasingly fragmented and anomalous and giving way to something new. Renewal and disintegration are both parts of the same process of transformation. But transformation is slow and gradual. As we aspire to play a meaningful role in this process, each of us must begin taking purposeful steps in the direction we want to see for our future without being overly preoccupied with immediate results.

Morris Berman's concept of the New Monastic individuals is a model of this purposive action. People can and do take action that is authentic and grounded in the belief that by its very authenticity it is a step toward real change. These are small steps that are unlikely to be heralded with tremendous fanfare as being "revolutionary." Rather, they are actions taken by individuals who have chosen to do things in ways that run against the current of what is expected in American life. These are the quiet renegades who make necessary and small changes that are positive, creative, and in a spirit of renewal. This should be done with care to preserve the treasures that are worth salvaging from the current declining system. These treasures are part of the basic elements that, when taken together with all that is new, will constitute the bricks and mortar of a new system—a new commonwealth. By being proactive in these small and individual ways—preserving and creating—each of us will be contributing to a great transformation that will bring a new system of economic production into being. But this can only happen when the time is right.

Recall Patrick Geddes's vision for civic regeneration based on ecological permanence and urban renewal. Geddes was hopeful that urban communities could evolve into living eutopias—social and economic systems that come into being when the conditions are conducive for them to do so. Geddes's vision of eutopia was given more contemporary and scientific expression in the 20th and 21st centuries with advanced ecological science and permaculture design systems. The core principles of permaculture are to meet the needs of people in ways that are fair, equitable, and ecologically permanent. We have to expand on these principles as we evolve, and with the vision of an emerging eutopia, we will establish an ideological basis for a livable, post carbon future. And, as

I have been arguing throughout these pages, as we develop new ideology, we also be crafting new institutions that will transform the rules of governance in our economy.

I also urge caution that we not expect to find the absolute solution in one institutional model. As we look for the treasures around us that provide hints or glimpses of the system to come, we only find fragments of a mysterious whole that is still in a state of becoming. Economic historian Karl Polanyi, in his book, *The Great Transformation: The Political and Economic Origins of Our Time* (1944), tells the story of how capitalism as a system emerged from preexisting institutional fragments preserved during the Renaissance and the rise of the secular nation-state.[65] There was no way people living in the Renaissance could foresee what kind of world these new capitalistic institutions would bring.

The same process will hold true as our future system emerges. We cannot see exactly what it will become, but we can contribute to its development nonetheless. The dynamics of the disintegration and renewal of economic systems tend to always follow the same pattern: as one crumbles on itself, it leaves behind vestiges that can be restored or recycled and used as parts that comprise the new system. The cohesiveness of this assemblage of parts is contextual as they tend to bond when the circumstances are right. Though seeing economies as total systems requires a holistic view, it is still necessary to understanding the substance or building blocks from which the systems are formed. On this point institutionalist economist Clarence Ayres observed that "Some sort of division of the social whole into parts is inevitable, and for this the familiar 'institutions' stand ready to hand."[66] What I describe in the chapter on the New Monastics are institutional alternatives. But we cannot project from these developments to see a wholly new economic system any more than we can understand a flower blossom by observing one petal in isolation. It is only when the bits and pieces are matched with other bits and pieces that the emergent properties of a wholly new system will be revealed. And, again, that can only happen when the time is right. The current peak oil moment is that time.

Elinor Ostrom teaches us that institutions are the rule-structured situations that we humans need for survival. Humans are institutional creatures and our lives are structured around the dos and don'ts of social behavior. Whether they are the rules of law, the bylaws of corporate governance, the shared strategies of a self-defined community, regulations in sports, or the rules parents set for their children, we live according to such strictures because human life would be impossible without them. Economic institutions are a subcategory of these rule-structured situations that are specifically for ordering the way we fashion a living out of the common pool of resources around us. Like everything else in the universe, however, these too are in a constant state of flux and transformation. It is here with creating new rule-structured economic environments that our real work of conscious transformation needs to be done. In multitudes of small, tucked-away places this work *is* being done and is gaining momentum and scope with each new initiative.

As this process of transformation gains momentum there will be resistance. People in general do not like it when the rules change or when they are forced to change their habits because a new rule-structured situation is being imposed on them. But in the bigger picture, this resistance is based on the delusion of permanence and the misconception that our future economic system is survivable in its current form. Resistance also comes from the fact that for a very long time Americans have been raised to believe that their main purpose in life is to chase after a largely illusory dream of a good life—the American Dream. In this dream, our happiness is supposed to derive from economic accomplishments: dwelling in taller skyscrapers, driving cars that can go anywhere and are equipped with everything imaginable, vacation property, a new closet full of clothes for every passing season, more electronic gadgets loaded down with more applications, and the rest of a long list that signifies material success. Who can blame someone for coveting these accoutrements of the good life when this has been the American way since the country's inception?

In his most recent book, *Why America Failed: The Roots of Imperial Decline* (2012), Morris Berman tells us how from the outset, America as nation and culture was never intended to be a truly democratic commonwealth. The roots of failure and decline according to Berman lie in what Samuel Adams described as ". . . the Rage for Profit and Commerce [*sic*]." America's economic institutions were crafted around the deeply rooted impulse for covetousness that was so reviled by Randolph Bourne and the other Young Americans as well as Schumacher. This superseded the more transcendent project of building a commonwealth genuinely crafted to serve the wellbeing of the population. Berman remarks ". . . material acquisition and technological innovation were druglike substitutes for a commonwealth, a truly human way of life, that Americans had largely rejected from early on."[67]

Since the very beginning, the vision of the United States as a nation and culture has between caught between a land made for unrepentant acquisitiveness on the one hand, and a land for a truly democratic commonwealth on the other. The vision of acquisitiveness won out and greed and materialism continue to be normalized in American culture. From our history we have inherited what is now a grossly unsustainable society in which, according to Berman, ". . . cash becomes the end-all for everybody."[68]

Perhaps many of us feel this emptiness and sense that we need to alter our course, yet find making such changes nearly impossible. Why is it so difficult to change? For one, this would require that Americans in significant numbers question the values of their own culture. For another, Americans' peculiar habit of always running headlong into the grab for riches has become so deeply ingrained and normalized that it has become a culturally determined addiction. It is very hard for addicts to imagine themselves free from their addictions, particularly ones that have been around for over two centuries. On this, Berman writes:

> If the American Dream is really about unlimited abundance, and if we are addicted to that as a goal, then alternatives to that way of life are simply too scary to contemplate. Try telling a full-blown alcoholic to put down that glass of Scotch . . . addiction has a certain "systemic" pattern to it that is typically not self-corrective. Both capitalism and alcoholism are characterized by cycles of increasing dysfunction, "runaway," and breakdown, and the system can do this for a fairly long time.[69]

Many of us have developed the belief that we can continue this forever simply because as a nation we have been doing this for a very long time. It would be bitter medicine for us to see our consumerist lifestyles in another way. But if we could swallow it, we would eventually see that such an addiction, like all addictions, is taking us down a path toward self-destruction. On this process of extending addictions to a social and cultural level, Buddhist sociologist Ken Jones writes:

> As it is with individual lives, so it is with institutions, societies, and cultures: They may be swept into ruin by karmic and other tangled conditionality even though they have the objective means to avert their fate and more than enough warning of it. The actors are driven

by addictive behavior and a kind of tunnel vision that is ultimately self-destructive. And when the majority is locked into mutually affirming karma it may be particularly difficult for even a well-informed minority to achieve a change of direction.[70]

This karmic trap forms a perfect circle of pathology. Our addiction to consumerism leaves us socially and spiritually malnourished, and with a weakened moral center. In a vain attempt to fill that emptiness we turn to more consumerism, which compels us to grasp for more money. The more habitual becomes our grasping, the more we feel the need to have things to show for the effort, which leads to more acquisitiveness. We'll keep lunging in circles like this, like a dog snapping at its own tail, until we collapse from exhaustion. Berman notes, "You really can't shoot heroin forever."[71] But we collectively live with the soothing hallucination that we can as long as there is technological "progress."

I think it is fair to say that most Americans would rather believe in the most dreamlike science fiction about how technology will save us than face the reality that our way of life cannot last. But as I mentioned above, this is attitude is largely justified with the fatalistic contention that people cannot change. Such fatalism is a death sentence on any positive movement for real change, and is derived, in part, from ideological indoctrination and the belief that the lure of big money is the only way people can be motivated to be productive. I say indoctrination here because capitalism depends on the perpetuation of this belief. For some people it may be true that their only motivation is money, but for many others it is not. Nonetheless, the debate on human nature, what motivates us, and whether or not we can be motivated by something other than the lure of riches is largely irrelevant because the idea of ever-ascending material standards of living is soon to be obsolete. Eventually people will change their habits, their belief in technology and their view of money because the end of Oil Age will force them to.

The idolatry of money and the obsession with the ever-ascending trajectory of economic growth are one and the same phenomenon. Money grows in the financial system where banks charge and pay interest. This means compounding financial returns, and those returns drive material economic growth because those returns can only be realized with product market expansion. If the ability of markets to expand is hitting the wall because of resource limitations, then our economy will begin exhausting its saved capital, which as of now is paper thin. Like computer software, household appliances, and most everything else associated with our system of mass production and mass consumption, the system itself was not built to last.

If the system overall can no longer expand but its driving institutions remain intact, then trouble is certain to follow. One segment of the population still driven by the will to expand will cannibalize the resources of another. This can only lead to endless conflict over dwindling resources. In fact, such conflicts are already happening and are primarily characterized by the already wealthy and powerful using military force against the already poor and weak. Take your pick as to how that force will be justified in the media and altogether too easily accepted by the American public. Moral objections aside, if we go down this path then this will only hasten the time at which we reach complete resource exhaustion. At that point the idolatry of money and growth obsession will end in its own violent death of obsolescence after all, only then we will have tragically destroyed any hope of creating a better future.

Of course it does not have to be this way—if we uproot our entrenched and unsustainable habits of thought and action. I certainly am not the first person to suggest this as a necessity for our long term survival. Many others for decades now have been imploring citizens and governments to approach these challenges presented by resource depletion openly and with positive energy. Since the shift to a post carbon economy is inevitable, why shouldn't we envision it changing into something more positive: just, ecologically sound, stable, vibrant, elegant, and healthy. But, again, to accomplish this we will have to create wholly new institutions and salvage the fragments of the ones that currently dominate our political and economic system. And all of this needs to be based on values, principles, and a new vision for our future.

To this end, let us revive the work of the meta-economists. For over a century these visionaries held the conviction that spiritual insight, aesthetic form, economic production, and community are inseparable facets of our humanity. They made a compelling case for the integration of artistic expression and human labor in such a way that work would have much deeper meaning and significance than a mere job. To see our work in this way is to develop all our inherent capabilities in order to live full, joyful lives. Humanizing economic production would have to be predicated on shifting its emphasis from endless growth and money accumulation to a much stronger commitment to developing our potentialities. This commitment would also have to be preserved in our institutions and integrated into new rules of governance.

They were also inspired by a romantic view of organic form and the beauty of nature. They did not see humans trying to mimic nature as do many advocates of "green capitalism," but rather sought to integrate what is naturally and uniquely human—our institutional and cultural nature—into a healthier ecological relationship with nature, and with a particularly urban flavor. There is no doubt that the societies of the world are going to be much more clustered into urban hubs as proximity becomes the most important feature to define post oil human habitats. Planning our living spaces in this way cannot happen without developing a deeper sense of community and an entirely distinct mode of economic production. Though writing in the 19th and 20th centuries, these visionaries created a framework for a future economic transformation away from the ravages of unstable global systems toward a network of decentralized economies. Each of these more localized economies is a network of separate institutions setting the rules for production and cooperation in ways that balance urban culture with the aesthetic and healthful benefits of nature.

Such work could also be seen as a political movement. For Bourne, Brooks, Frank, and Mumford, local urban cultures were the ground zero for political transformation. Though they were concerned, as Berman expresses, that America always been a place of hustling for cash and being trapped in a consumer culture as the object of living, they held onto the hope that people could nonetheless aspire to a more fulfilling life. The Young Americans, especially, sought to build a counterforce to popular culture that had become dominated by consumerism, money grubbing, and meaningless fads. Like their British predecessors they found consolation in community-based reforms centered on the development of humanity's creative, productive, and spiritual potentialities. And like John Dewey, they saw that these potentialities could blossom into full development with civic activism and participation. Unlike Dewey, however, the Young Americans held that such activism should be directed not at national party politics, but at community life.

As we explore the development of community-based economic alternatives in the spirit of these visionaries, we are looking into something much deeper than mere "going local" business models. Theirs was a vision of cultural renewal or transformation such that the daily business of production and consumption is a means to a more transcendent end. The end is human social and cultural evolution. Evolving toward something better. For them, the potential of each individual remains stunted in a society that offers little more than vague opportunities for material gratification and the hustle for profit. By recreating community life through integrating the life-giving qualities of nature with more refined cultural expression through crafts, the local community and economy can become a rich environment within which our species can develop to its full potential.

For Schumacher, this developmental process cannot happen without insight and wisdom. And though he wrote passionately about the need for appropriate forms of technology and new models for economic development, Schumacher emphasized that these are eclipsed by the much larger challenge of finding guidance through a realization of higher-level truths and wisdom. The crises we face stem not only from how we act in the world, but also from how we perceive the world.

Bringing this vision into action and real change is no small matter and will be a process that will take time and enormous effort. It may even appear impossible from where we are standing now, but I would venture to say that it does not seem as impossible as trying to survive another few decades of economics as we know it. To quote Howard again, our economic system is "... the best which a society largely based on selfishness and rapacity could construct ..." But for all its glitter, this system is on the descent, and like all large-scale transformations in history, its disintegration is giving way to something new. As agents for change, our job is to shape this new system in the best way we can.

Once we free ourselves from the mindset and expectations associated with this collapsing system, it will become much easier for us to imagine ourselves evolving into an economically and ecologically healthier society. When we are no longer burdened by the cultural baggage associated with a system that is psychotically trying to achieve infinite growth on a finite planet, we can more effectively explore the possibilities for consciously evolving toward new and healthier ways of living. The meta-economists from Ruskin to Schumacher, the original architects of permaculture, and the transitioneers have given us a rich body of ideas from which we can build a new ideology of change. That is, we can fundamentally change the way we approach change.

But ideas and ideological transformation will only get us half way. To complete the journey, we must be actively getting our hands dirty in our communities in ways that are consistent with this ideology. The agents for change, like Berman's New Monastic individuals, are the people who are actively setting out to create new models for organizing economic activity. But these models are not fashioned out of thin air, they are created out of the bits and pieces of creative efforts that are underway all around us. These pieces are like the individual genes that make up the DNA which will shape the overall systemic whole. It is with this idea that I present the various New Monastic initiatives.

The case studies I presented here are those that have some promise to provide the institutional raw material from which a more comprehensive system can evolve. Each in one way or another has some affinity with the vision inspired by our meta-economists, and each element that glows with this affinity can be brought together with every other element to give rise to an emergent entity—a new commonwealth.

One element is the Schumacherian concept of smaller scale production systems through localization and local ownership either through worker ownership or community centered cooperatives. We have looked at a number of ways this can be achieved by developing business models that are worker-owned such as the New Belgium Brewery, Namasté Solar, and the Evergreen Cooperatives. By empowering the people who actually do the work with an ownership stake in the company, the business is automatically set loose from the dominant model of giant publicly traded corporations and their disinterested class of shareholders whose only concern is generating returns and growth. Locally owned, family-owned, or worker-owned businesses, even if they are for-profit enterprises, do not have to provide returns as such for investors. Some may want this kind of growth, but it is not institutionally required as a fiduciary responsibility nor profit merely for profit's sake.

Recall, however, that going local is not enough to carry the mantle of economic survival in a post carbon world. The meta-economists' reverence for natural beauty and ecological responsibility can inspire us to go beyond the limited but important elements of genuine commitment to renewable energy and resource stewardship on the business side, and reinforce a shift away from runaway consumerism. Their adulation of craft traditions can also inspire the development of craft institutions and support of artisans like those at Resource Revival who transform worn-out bicycle parts into clocks, furniture, bottle openers, and other things that are useful and exemplify the personal joy that comes from true craftsmanship and the integration of artistic creativity with economic production.

If we, the change agents for the future, were to refashion our communities around these elements of localization, renewable energy, resource conservation and stewardship, and a celebration of craft traditions, then these communities would look radically different from what currently exists. But building a new commonwealth stemming from our meta-economic vision goes much further. If all the local businesses and households became members of financial cooperatives or opened accounts at community banks that were organized like the Permaculture Credit Union with the emphasis that the financial resources stay in the community, then again things would look very different. And they would look even more different if, as people paid their taxes, the state treasuries were to deposit their revenues in banks patterned after the Bank of North Dakota. These funds could be used to help finance all kinds of tremendous economic development projects consistent with our core values. They could also provide additional resources for land conservation projects as are being carried out by the Santa Fe Farmers Market Institute. Each community or state could build its own health care system patterned after Vermont's initiative or the Freelancers Union. There would be a greater difference still if every community passed home rule legislation as CELDF has done in order help protect the rights of communities and their habitats from corporate power. Every community would indeed look quite different if it had city planners like those in Portland, Oregon, who are committed to creating comfortable, livable, sustainable urban environments.

And these make up just the beginning. If we took the meta-economic vision to heart and set out to rebuild our community-based economic institutions incorporating all these different elements, we would in fact be fostering the evolution of an entirely new system. So new, in fact, that we don't yet even have an -ism for it. But there is no need to be overly preoccupied with what this system will be called or what it will look like. What is most important is that it be fashioned out of genuine care, wisdom, values, and insight. If it is, it cannot go wrong. For this transformation to take place, and there is no persuasive reason why it could not, the emphasis needs to be placed on rewriting the rules of the game. The more we take this kind of action, the more we change our perceptions of the world around us, and the more our perceptions of the world around us change, the more our actions will shift accordingly.

This, I believe, is what Lewis Mumford saw for our future. Mumford's writing career spanned many decades up to the 1960s and overlapped with Schumacher's in the 1970s. In *The City in History* (1961), his two-volume *Myth of the Machine: Technics and Human Development* (1967) and *The Pentagon of Power* (1970) his ideas were prescient and, like those of Schumacher, are perhaps more relevant now than they were in their time. Mumford's conclusions are summarized by Casey Nelson Blake:

Mumford had developed a more consistent and devastating perspective on modern civilization, counterposing political localism, respect for handicrafts and decentralized technical systems, and an ethos of personal restraint to the religion of bigness, technological innovation, and state power he saw everywhere at work in postwar America.[72]

Mumford developed a consistent theme throughout the span of his working life: envisaging a kind of cultural retreat from business and politics as usual that sought "... not to capture the citadel of power, but to withdraw from it and quietly paralyze it." Such a shift is as much spiritual as it is political, such that this withdrawal can "restore power and confident authority to its proper source: the human personality and the small face-to-face community."[73]

Thus our work is cut out for us. Let us raise new generations and build new communities of change agents to deal effectively with the coming problems we will undoubtedly face when the world begins to run out of oil: to reform education, to create new models of health care, new financial cooperatives and community banks, state banks, B Corporations, households, and scores of other institutions that are governed by new rules formed out of core values derived from insight and wisdom. All of these initiatives will eventually wedge themselves against what is now a crumbling and antiquated global system.

NOTES

1. *NPG Special Report*, "Forgotten Fundamentals of the Energy Crisis," 1977.
2. Noted in paper Pimentel presented at the 1977 annual meeting of AAPT-APS, Chicago, 1977, cited in the *NPG Special Report*.
3. Kenneth Deffeyes, *Beyond Oil: The View From Hubbert's Peak* (New York: Hill and Wang, 2005), 3.
4. Energy Information Administration, "International Petroleum Monthly," January 2011.
5. Jeff Rubin, *Why Your World is About to Get A Whole Lot Smaller: Oil and the End of Globalization* (New York: Random House, 2009), 79.
6. Sandra Postel, *Last Oasis: Facing Water Scarcity* (New York: W. W. Norton, 1992), xiv.
7. IPCC Fourth Assessment Report: Climate Change 2007.
8. US Department of Energy Report to the Biological and Environmental Advisory Committee (BERAC), October 2010.
9. Graham Bell, *The Permaculture Way: Practical Steps to Creating a Self-Sustaining World* (White River Junction, VT: Chelsea Green, 2008), 22.
10. Helena Norberg-Hodge, "Beyond The Monoculture: Strengthening Local Culture, Economy And Knowledge," Countercurrents.org, April 3, 2010, http://www.countercurrents.org/hodge030410.htm.
11. Ken Jones, *The New Social Face of Buddhism: An Alternative Sociopolitical Perspective* (Boston: Wisdom Publication, 2003), 17.
12. Rob Hopkins, *The Transition Handbook: From Oil Dependency to Local Resilience* (White River Junction, VT: Chelsea Green, 2008), 141.
13. David Hamilton, *Evolutionary Economics: A Study of Change in Economic Thought* (Albuquerque NM: University of New Mexico Press, 1978), 77.
14. Herbert Marcuse, *One Dimensional Man: Studies in the Ideology of Advanced Industrial Society* (Boston: Beacon Press, 1964), 51.
15. Vaclav Havel, *Disturbing the Peace: A Conversation with Karel Huizdala* (New York: Vintage, 1991), 14.
16. Jones, *The New Social Face of Buddhism*, 72.
17. Ibid.
18. David R. Loy, *Money, Sex, War, Karma: Notes for a Buddhist Revolution* (Boston: Wisdom Publication, 2008), 89.
19. Sulak Sivaraksa, *Seeds of Peace: A Buddhist Vision for Renewing Society* (Berkeley: Parallax Press, 1992), 8.
20. Quoted in David Owen, "The Inventor's Dilemma," *The New Yorker*, May 17, 2010, 47.
21. James Glanz, "Power, Pollution, and the Internet," *New York Times*, September 23, 2012.

22. Juliet Schor, *Plenitude: The Economics of True Wealth* (New York: The Penguin Press, 2010), 28–29.

23. Jeane and Dick Roy, "Deep Sustainability" in *Earth Matters: The Newsletter of the Northwest Earth Institute*, 12, No. 1, 1–2.

24. Woody Tasch, *Slow Money: Investing as if Food, Farms, and Fertility Mattered* (White River Junction, VT: Chelsea Green, 2008), 48.

25. "Durban Declaration on Carbon Trading," South Africa, 2004.

26. Kris De Decker, "Leave the Algae Alone," *Low Tech Magazine*, April 4, 2008, http://www.low-techmagazine.com/2008/04/algae-fuel-biof.html.

27. "World Energy Outlook Factsheet," International Energy Agency, 1–6, http://www.worldenergyoutlook.org/media/weowebsite/2012/factsheets.pdf.

28. "Press Release," IPCC, May 2011, http://www.ipcc.ch/news_and_events/docs/ipcc33/IPCC_Press_Release_11612_en.pdf.

29. Ted Trainer, *Renewable Energy Cannot Sustain a Consumer Society* (London: Springer, 2010), 42.

30. Doug Henwood, "How to Learn Nothing From Crisis," *Left Business Observer*, 125, February 2010.

31. E. F. Schumacher, *Small Is Beautiful: Economics as if People Mattered* (New York: Harper and Row, 1973), 101.

32. Ibid., 93.

33. Randolph Bourne, "Trans-national America" in Olaf Hansen, *The Radical Will: Randolph Borne Selected Writings, 1911–1918* (New York: Urizen Books, 1977), 264.

34. John Ruskin, *The Stones of Venice*, J. G. Links ed. (New York: DeCapo Press, 1960), 244.

35. P. D. Anthony, *John Ruskin's Labour: A Study of Ruskin's Social Theory*, (Cambridge, UK: Cambridge University Press, 1983), 54.

36. Ebenezer Howard, *Garden Cities of To-morrow* (Harvard University Press reprint of the original published by Swan Sonnenschein, London, 1902), 20.

37. Ibid., 22.

38. Ibid. (reprint of the 1898 edition by MIT Press, 1965), 57.

39. Ibid., 29.

40. Brooks, *America's Coming-of-Age* (New York: Doubleday, 1958), 63, 68.

41. Bourne, "Trans-national America," 254–55.

42. Ibid.

43. Lewis Mumford, *The Story of Utopias* (New York: Viking Press, 1962), 209, 212.

44. Waldo Frank, *Re-discovery of American: An Introduction to a Philosophy of American Life* (New York: Charles Scribner's Sons, 1929), 70; "The Treason of the Intellectuals," Modern Quarterly, 5 (Spring, 1929), 163; "In Defence of Our Vulgarity," in *In the American Jungle, 1925–1936* (New York : Farrar and Rinehart, 1937), 111.

45. Lewis Mumford, *Findings and Keepings: Analects for an Autobiography* (New York: Harcourt, Brace, Jovanovich, 1975), 206.

46. Mumford, "A Search for the True Community," in *The Menorah Treasury: Harvest of Half a Century*, ed., Leo W. Schwartz (Philadelphia: Jewish Publication Society of America, 1964), 859.

47. Schumacher, *A Guide for the Perplexed* (New York: Harper and Row, 1977), 48.

48. Schumacher, *Small Is Beautiful*, 145.

49. Ibid., 175.

50. Howard, *Garden Cities of To-morrow*, 122–123.

51. Eric Foner, *A Short History of Reconstruction, 1863–1877* (New York: Harper and Row, 1990), 41.

52. Morris Berman, *The Twilight of American Culture* (New York: W. W. Norton, 2000), 9.

53. Ibid., 10.

54. Elinor Ostrom, *Understanding Institutional Diversity* (Princeton, NJ: Princeton University Press, 2005), 186.

55. "15 Green Cities," *Grist Magazine*, http://grist.org/article/cities3/.

56. Richard Gibert and Anthony Perl, *Transport Revolutions: Moving People and Freight Without Oil*, 2nd ed. (Gabriola Island, Canada: New Society Publishers, 2010), 2–3.

57. Paul Sherrer, "The 2000 Watt Society—Standard or Guidepost?" *Energy-Speigel, Facts for the Energy Decisions of Tomorrow*, no. 18 (2007).

58. Nel Noddings, "Global Citizenship, Promises and Problems," Nel Nodding, ed., *Educating Citizens for Global Awareness* (New York: Teachers College Press, 2005), 3.

59. Ibid.

60. Noddings, "Global Citizenship, Promises and Problems," 79.

61. Alpha C. Chiang and Kevin Wainwright, *Fundamental Methods of Mathematical Economics*, 4th ed. (New York: McGraw-Hill/Irwin, 2004), 332.

62. Philip Mirowski, *Against Mechanism* (Lanham, MD: Rowman and Littlefield, 1988), 115.

63. E. K. Hunt, *Property and Prophets: The Evolution of Economic Institutions and Ideologies* (New York: ME Sharpe, 2003), 126.

64. Rita Mae Brown, *Sudden Death* (New York: Bantam, 1983), 83.

65. Karl Polanyi, *The Great Transformation: The Political and Economic Origins of Our Time* (Boston: Beacon Press, 1944).

66. Clarence Ayres, *The Theory of Economic Progress* 3rd ed. (Kalamazoo, MI: New Issues Press, 1978), 178.

67. Morris Berman, *Why America Failed: The Roots of Imperial Decline* (New York: Wiley, 2011), 24.

68. Ibid., 55.

69. Ibid., 66.

70. Ken Jones, *The New Social Face of Buddhism: An Alternative Sociopolitical Perspective* (Boston: Wisdom Publications, 2003), 27.

71. Berman, *Why America Failed*, 70.

72. Casey Nelson Blake, *Beloved Community: The Cultural Criticism of Randolph Bourne, Van Wyck Brooks, Waldo Frank, and Lewis Mumford* (Chapel Hill, NC: University of North Carolina Press, 1990), 287.

73. Ibid.

INDEX

About the Author

JOEL MAGNUSON is an independent economist based in Portland, Oregon, USA. He is a visiting fellow at the Ashcroft International Business School at Anglia Ruskin University in Cambridge, England, serves as an international advisor to Anglia's journal *Interconnections*, and is on the faculty at the East West Sanctuary in Nagykovácsi, Hungary. He is the author of *Mindful Economics: How the US Economy Works, Why It Matters, and How It Could Be Different* (Seven Stories Press, 2008) as well numerous articles in journals and anthologies in the US, Europe, and Japan.

About Seven Stories Press

SEVEN STORIES PRESS is an independent book publisher based in New York City. We publish works of the imagination by such writers as Nelson Algren, Russell Banks, Octavia E. Butler, Ani DiFranco, Assia Djebar, Ariel Dorfman, Coco Fusco, Barry Gifford, Martha Long, Luis Negrón, Hwang Sok-yong, Lee Stringer, and Kurt Vonnegut, to name a few, together with political titles by voices of conscience, including Subhankar Banerjee, the Boston Women's Health Collective, Noam Chomsky, Angela Y. Davis, Human Rights Watch, Derrick Jensen, Ralph Nader, Loretta Napoleoni, Gary Null, Greg Palast, Project Censored, Barbara Seaman, Alice Walker, Gary Webb, and Howard Zinn, among many others. Seven Stories Press believes publishers have a special responsibility to defend free speech and human rights, and to celebrate the gifts of the human imagination, wherever we can. In 2012 we launched Triangle Square *books for young readers* with strong social justice and narrative components, telling personal stories of courage and commitment. For additional information, visit www.sevenstories.com.